Preface

In this edition the opportunity has been taken to update the contents to keep abreast of developments in the field of digital electronics. This work contains two new chapters, the first (chapter 10) dealing with alphanumeric displays and associated circuits, whilst the second (chapter 11) covers both digital-to-analogue convertors and analogue-to-digital convertors. In addition, the work in the other chapters has been revised, and dated material has been removed. It is felt that the changes will enhance the book both as a digital logic course book and for background reading in other fields. The chapter on fluid logic which was in the enlarged first edition has been deleted, and is available in expanded form in the companion book *An Introduction to Fluid Logic* (McGraw-Hill).

The logic symbols adopted in the first edition have been retained in this book. The advantages gained by this are twofold, namely that the function generated by each logic gate is clearly stated on the gate, thereby making logic diagrams easily understood, and that there is a significant saving in cost to the purchaser of the book by avoiding the necessity of re-drawing many of the circuits. For the convenience of readers, a range of British, American, and European logic symbols is provided in the appendix at the end of the book.

I would like to take this opportunity to express my gratitude to the McGraw-Hill reviewers for their constructive comments on the revised edition. I am also grateful for the many comments I have received from time-to-time from my colleagues at the North Staffordshire Polytechnic, and also from associates in the teaching profession and in industry. Thanks are also due to the McGraw-Hill production and editoral staff for maintaining the quality of the book.

Once again I am indebted to my wife for the help and assistance she has given during the preparation of this edition.

Noel M. Morris
Meir Heath

Preface to the first edition

Today, engineers are meeting a new industrial revolution. Rapid developments in the field of logical devices have led to the application of semiconductor and other switching devices, to the control of engineering systems. The advent of integrated circuits has made ever-greater demands on the ingenuity of the engineer in utilizing them to the best effect.

To aid the understanding of logical problems, the approach used in this book is systematic, commencing with an introduction to numbers and numbering systems before progressing to the fundamentals of logic. In the fourth chapter, the process of converting logical equations into logical hardware is described, together with techniques for the minimization of the number of logic gates used.

Before progressing to sequential networks, the rudiments of basic flip-flops are analysed. One chapter is devoted to asynchronous counters, and one to synchronous counters, so important is their place in engineering systems. Up to this point in the book, a 'systems' approach is used, so that the restrictions imposed by specific devices do not inhibit discussion of the problems.

Chapters eight and nine are devoted to discussion of electronic circuit principles. Device theory has been limited to that necessary for an understanding of the principles involved. With the advent of integrated circuits, detailed design principles of logical devices are now less meaningful. The principles of modern film and monolithic integrated circuits are fully described.

In writing this book, I have endeavoured to select topics which are of interest to designers, users, and students of engineering systems, rather than to orient it toward computer design. I hope that engineers who have not hitherto been concerned with logic circuits will find something of interest in the book. The text is illustrated throughout by worked examples, and a bibliography is provided for reading matter on related topics.

The book should be of value to all students of logic, be they graduates or undergraduates. It is hoped that the book will prove valuable to students in H.N.D., H.N.C., and City and Guilds courses.

I would like to thank manufacturers and users of logic circuits for the advice and assistance so freely given, and the British Standards Institution, the American Institute of Electrical and Electronics Engineers and the British Compressed Air Society for the information on standard graphic symbols which are incorporated in the appendix. I am also indebted to my colleagues at the North Staffordshire Polytechnic for the many useful discussions that took place during the preparation of the book, and to my wife for the valuable assistance she has rendered in checking and typing the manuscript.

N. M. Morris

Contents

List of symbols

I_C	Collector current.
I_B	Base current.
I_{CE0}	Leakage current in the common emitter mode with $I_B = 0$.
h_{fe}	Small-signal forward current transfer ratio.
h_{FE}	Static value of the forward current transfer ratio.
$h_{FE(sat)}$	The value of h_{FE} in the saturated mode.
V_{CC}	Collector circuit supply voltage.
V_{CE}	Collector-to-emitter voltage.
$V_{CE(sat)}$	Collector-to-emitter voltage in the saturated mode.
V_{BB}	Base circuit bias voltage.
V_B	Base circuit voltage.
$V_{BE(sat)}$	Base-to-emitter voltage in the saturated mode.
V_h	Hysteresis voltage.
V_T	Threshold voltage.
E_D	Diode 'catching' voltage.
E_S	A supply voltage.
R_B	Resistance in the base circuit.
R_C	Resistance in the collector circuit.
$r_{CE(sat)}$	Transistor saturation resistance.
r, R	Radix of numbering systems.
T_j	Junction temperature.
t_r	Rise time.
t_d	Turn-on delay.
t_s	Storage time.
t_f	Fall-time.
ω_β	Cut-off frequency.
M	Fan-out.
N	Fan-in.

1. Numbers and numbering systems

1.1 Terminology

The most common numbering system in use today is the *decimal system* or *denary system* which utilizes ten digits 0, 1, 2, 3, 4, 5, 6, 7, 8, 9. The number of digits used in the system is known as its *base* or *radix*; the radix of the decimal system is ten. Another common system, known as the *octal* system, having eight digits 0, 1, 2, 3, 4, 5, 6, 7 is used in computer systems. The radix of the octal system is eight.

Each term in a number is associated with the radix raised to a power according to its position in the number, i.e., the decimal number 9 is in the zero-order position, while decimal 91 has a 9 in the first-order position, and a 1 in the zero-order position as follows

$$\text{decimal } 9 = 9 \times 10^0$$
$$\text{decimal } 91 = (9 \times 10^1) + (1 \times 10^0)$$
$$\text{octal } 21 = (2 \times 8^1) + (1 \times 8^0)$$

The space to both the left and right of any number is understood to be filled with zeros:

$$\text{decimal } 7 = 07 \cdot 0 = 00007 \cdot 0000, \text{ etc.}$$

When adding numbers manually, the position of the decimal point is clearly understood, and not all the zeros are written down, but in a computer the number must be specified exactly, and information about the position of the decimal point must be given.

In any natural system of numbers the representation of zero is $000 \cdot 000$, and that for unity is $001 \cdot 000$. Successive numbers are not always the same in all systems. In some numbering systems the number $000 \cdot 000$ does not exist, although all systems have some combination of numbers to represent zero.

1

Numbers are generally expressed in the form of a polynomial as follows

$$N = d_n r^n + d_{n-1} r^{n-1} + \ldots + d_2 r^2 + d_0 r^0 + d_{-1} r^{-1} + \ldots + d_{-n} r^{-n}$$

where r is the radix of the system, and d is an integer whose value lies in the range $0 < d < (r-1)$. In the decimal system, the value of d lies in the range zero to nine.

The simplest system of all, the *binary* system has only two values, 0 and 1. Binary digits are often referred to as 'bits' (*bi*nary dig*its*), a ten-bit number having ten binary digits.

The *duodecimal* system uses twelve units 0, 1, 2, 3, 4, 5, 6, 7, 8, 9, t, e. The last two numbers, ten and eleven, are usually given in their decimal form 10 and 11, respectively. These are not true duodecimal numbers, but are *decimal coded duodecimal*. For convenience, when discussing numbers, the radix is written as a suffix, viz.:

$$\text{decimal } 912 = 912_{10}$$
$$\text{binary } 1101 = 1101_2.$$

Table 1.1

System	Binary	Octal	Decimal	Duodecimal	Hexadecimal
Radix	2	8	10	12	16
	0	0	0	0	0
	1	1	1	1	1
	10	2	2	2	2
	11	3	3	3	3
	100	4	4	4	4
	101	5	5	5	5
	110	6	6	6	6
	111	7	7	7	7
	1000	10	8	8	8
	1001	11	9	9	9
	1010	12	10	t	a
	1011	13	11	e	b
	1100	14	12	10	c
	1101	15	13	11	d
	1110	16	14	12	e
	1111	17	15	13	f
	10000	20	16	14	10
	10001	21	17	15	11
	10010	22	18	16	12
	10011	23	19	17	13
	10100	24	20	18	14
	10101	25	21	19	15
	10110	26	22	1t	16
	10111	27	23	1e	17
	11000	30	24	20	18

A simple conversion between duodecimal and decimal, illustrating the effect of coding one number in terms of the other, follows.

$$t_{12} = (0 \times 12^1) + (t \times 12^0) = (1 \times 10^1) + (0 \times 10^0) = 10_{10}$$
$$e_{12} = (0 \times 12^1) + (e \times 12^0) = (1 \times 10^1) + (1 \times 10^0) = 11_{10}$$
$$10_{12} = (1 \times 12^1) + (0 \times 12^0) = (1 \times 10^1) + (2 \times 10^0) = 12_{10}.$$

Another popular number system, known as the *hexadecimal system*, has a radix of sixteen and uses the values 0, 1, 2, 3, 4, 5, 6, 7, 8, 9, a, b, c, d, e, and f to represent the numbers zero to fifteen, respectively. The first twenty-four numbers and their representative values in several codes are given in Table 1.1.

1.2 Converting an integer of any radix to its decimal equivalent

To convert any integer into the equivalent decimal number, expand the integer as a polynomial in powers of the radix and add the terms.

$$1101_2 = (1 \times 2^3) + (1 \times 2^2) + (0 \times 2^1) + (1 \times 2^0)$$
$$= (8 + 4 + 0 + 1)_{10} = 13_{10}$$
$$1e2t_{12} = (1 \times 12^3) + (e \times 12^2) + (2 \times 12^1) + (t \times 12^0)$$
$$= (1728 + 1584 + 24 + 10)_{10} = 3346_{10}$$

1.3 Converting a decimal integer to its equivalent in any radix

Divide the integer repeatedly by the radix, successive remainders giving the required number. To convert 200_{10} to radix 8 the procedure is as follows.

$$8 \,)\, \underline{200}$$

$$8 \,)\, \underline{25} \quad \text{remainder 0 (least significant digit)}$$

$$8 \,)\, \underline{3} \quad \text{remainder 1}$$

$$0 \quad \text{remainder 3 (most significant digit)}$$
$$200_{10} = 310_8$$

This process can be used to convert an integer in any radix to an integer in any other radix, but the division must be carried out in the new radix. Since we are acclimatized to division in the decimal system, it is appropriate to translate from the decimal system when carrying out long division.

1.4 Converting a binary integer to an octal number

The binary number must be divided into groups of three, commencing with the 2^0 digit. Each group of three is then written down as its decimal equivalent.

$$1101111_2 = 001,101,111_2 = 157_8$$

1.5   Representation of a number less than unity

In practice quantities smaller than unity need to be recorded. These quantities
are indicated by placing a dot or *radix point* before the fractional part of the
number. The 'point' indicates where the power to which the radix is raised
becomes negative. In the decimal system it is referred to as the decimal point, in
the binary system as the binary point, etc. Some examples are given below.

$$0 \cdot 25_{10} = (2 \times 10^{-1}) + (5 \times 10^{-2})$$
$$3 \cdot 42_5 = (3 \times 5^0) + (4 \times 5^{-1}) + (2 \times 5^{-2})$$
$$1 \cdot 11_2 = (1 \times 2^0) + (1 \times 2^{-1}) + (1 \times 2^{-2})$$

To convert any positive number in any radix to the decimal system, expand
the number in powers of the radix and add the terms.

When converting a decimal number to an equivalent number in any radix, the
parts to the left and right of the decimal point must be dealt with separately.
The part to the left, the integral part, is dealt with by the method outlined in
section 1.3. The fractional part is dealt with by multiplying it repeatedly by the
radix, the resulting integral parts giving the required number. To convert decimal
$46 \cdot 375$ to binary the procedure is as follows.

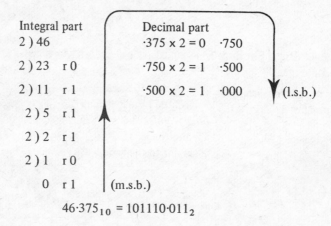

Integral part

2) 46

2) 23   r 0

2) 11   r 1

2) 5   r 1

2) 2   r 1

2) 1   r 0

0   r 1   (m.s.b.)

Decimal part

$\cdot 375 \times 2 = 0$ $\cdot 750$

$\cdot 750 \times 2 = 1$ $\cdot 500$

$\cdot 500 \times 2 = 1$ $\cdot 000$ (l.s.b.)

$$46 \cdot 375_{10} = 101110 \cdot 011_2$$

1.6   Number length

A system with a large radix is capable of conveying a large number using only a
few digits. Thus the duodecimal system is more economic than either the
decimal or binary systems, but it is not always convenient to work with in terms
of arithmetic.

To obtain an approximate relationship between the length, or number of
digits used, of numbers in two systems of radix 'r' and 'R' respectively, let

m = a '1' in the mth position of the number of radix r
n = a '1' in the nth position of the number of radix R

If

$$1.r^m \simeq 1.R^n$$

then

$$m \log_{10} r \simeq n \log_{10} R$$

or

$$m \simeq \frac{n \log_{10} R}{\log_{10} r}$$

Thus the approximate length of the binary equivalent of a decimal number of 'n' digits is $n \log_{10} 10/\log_{10} 2 \simeq 3\cdot33\,n$. Since a number comprises an integral number of digits the binary equivalent of a decimal number may require three to four times as many digits. For example, $9_{10} = 1001_2$, $54_{10} = 110110_2$.

1.7   Binary-decimal codes

In computers and logic systems, information and data is conveyed in binary form. The decimal number 24 can be conveyed as binary 11000. The latter number is in the *natural* or *pure binary code*.

The decimal system is the most convenient form of number representation for humans, whilst the binary system is the most convenient for logic devices. By *coding* decimal numbers in binary form, suitable numbering systems can be devised which are acceptable both to man and to the machine element.

Any decimal number can be represented by a group of four binary digits, as shown in Table 1.2. In the 8421 *binary-coded-decimal* (BCD) system, the least significant digit has a value $2^0 (= 1)$, and the most significant digit has a value $2^3 (= 8)$. In this respect the code is similar to the pure binary code.

Table 1.2

Binary-decimal codes

Decimal number	BCD codes				
	8421	2421	7421	642(−3)	5211
0	0000	0000	0000	0000	0000
1	0001	0001	0001	0101	0001
2	0010	0010	0010	0010	0011
3	0011	0011	0011	1001	0101
4	0100	0100	0100	0100	0111
5	0101	1011	0101	1011	1000
6	0110	1100	0110	0110	1001
7	0111	1101	1000	1101	1011
8	1000	1110	1001	1010	1101
9	1001	1111	1010	1111	1111

After a count of 9_{10} the 8421 BCD values or *weights* change by a factor equivalent to decimal ten. This is illustrated in Table 1.3. The first four bits of the code are used to convey decimal digits in the range zero to nine, while the next four bits convey the decimal numbers 10, 20, 30, . . ., 80, 90. In this way eight bits convey a maximum number of decimal 99.

Table 1.3

Binary-decimal codes

Decimal number	8421 BCD								2421 BCD							
	80	40	20	10	8	4	2	1	20	40	20	10	2	4	2	1
9	0	0	0	0	1	0	0	1	0	0	0	0	1	1	1	1
10	0	0	0	1	0	0	0	0	0	0	0	1	0	0	0	0
11	0	0	0	1	0	0	0	1	0	0	0	1	0	0	0	1
:																
79	0	1	1	1	1	0	0	1	1	1	0	1	1	1	1	1
80	1	0	0	0	0	0	0	0	1	1	1	0	0	0	0	0
:																
98	1	0	0	1	1	0	0	0	1	1	1	1	1	1	1	0
99	1	0	0	1	1	0	0	1	1	1	1	1	1	1	1	1

Many forms of code have been developed by altering the weighting of the four digits, but only a few of the many possible combinations have any practical value.

Although four binary bits can represent all ten decimal digits, it is possible for the four bits to represent a sequence of $2^4 = 16$ different states. Thus, in any 4-bit BCD code sequence there are six *redundant states* which, in the 8421 BCD code, are the states 1010 to 1111, inclusive. In fact, using 4 bits there are more than 29 000 million different codes available. Of these, only 70 are weighted codes, 17 having *positive weights*, including the 8421, the 2421, the 7421 and the 5211 BCD codes shown in Table 1.2. Some codes have *negative weights*, an example being the 642(−3) code in Table 1.2; in this code, a '1' in the least significant position of the code group represents a decimal value of −3.

Readers will also observe that there may be more than one method of representing a decimal number within a given code sequence. For example, in the 2421 BCD code the decimal number 4 may be represented in the form 0100 or 1010; in fact there are 32 different 2421 BCD codes. The 2421 BCD code listed in Table 1.2 is also described as a *self-complementing code*; self-complementing codes are useful in arithmetic systems, since they allow subtraction to be carried out by a process known as *complement addition*. The latter process is described more fully in section 2.2. In a self-complementing code if, in the code sequence representing number N, the 0's are changed into 1's and the 1's into 0's, then the resulting binary sequence is the same as that for the value $(9 - N)$. For example, the 2421 BCD representation of 3_{10} is 0011, and that for $(9 - 3)_{10} = 6_{10}$ is 1100; the latter binary sequence could have been obtained merely by converting the 0's into 1's and the 1's into 0's in the binary sequence for 3_{10} (the converse also applies).

1.8 Error detection

There are many possiblities for error in any calculation, not the least of which is human error in presenting information to the logic system.

In the 8421 BCD system the number 1111_2 is a forbidden combination since it represents decimal 15, the true representation of this number in 8421 BCD being 0001 0101. There are six *forbidden* or *can't happen* conditions in 8421 BCD, namely 1010, 1011, 1100, 1101, 1110, and 1111. If one of these is detected, it is due to an error of one kind or another.

In order to increase the reliability of the numbering system, redundancy is introduced and in general a high *redundancy* results in a reliable number system. An example of a highly redundant system is the English language. It si possible in thes sintence to detect and correct the three errors without difficulty.

The simplest method of error detection is to use an additional bit, known as a *parity bit*, for each group of the code. If *odd parity* is used, the parity bit (P in Table 1.4) is zero if the total decimal sum of the 1's in the group is odd, and it is unity if the sum of the 1's is even. *Even parity* can also be used, as shown in Table 1.4, odd parity having the advantage that no number is represented by a row of zeros. The parity bit may be written on either side of the group of bits. The parity of the group is checked against the parity bit during processing to investigate the possibility of an error having occurred. Any error detected in this way cannot be corrected, and a parity failure merely results in the machine or process involved being made to stop.

A useful concept in error checking is the *code distance*; the *minimum distance* or *Hamming distance*, d, of any code is the minimum number of bits

Table 1.4

Odd and even parity checks

Decimal number	Odd parity P	8421	Even parity P	8421
0	1	0000	0	0000
1	0	0001	1	0001
2	0	0010	1	0010
3	1	0011	0	0011
4	0	0100	1	0100
5	1	0101	0	0101
6	1	0110	0	0110
7	0	0111	1	0111
8	0	1000	1	1000
9	1	1001	0	1001

which change when the code group changes from any number to an adjacent number in the sequence. The minimum distance of the 8421 BCD code is 1 (which occurs when the code group changes from 0000 to 0001). *The number of errors which can be detected by a code of minimum distance d is (d − 1) or fewer.* Thus, in order to detect one error, the distance of the code must be at least two. Readers will note that the minimum distance of both codes in Table 1.4 is two.

1.9   Error correcting codes

There are classes of codes which not only indicate if an error exists, but can also be used to correct the error; such codes are described as *error correcting codes.*

Several parity bits are used in these codes, thereby increasing the redundancy of the numbering system; this feature allows code sequences to be checked for errors and, if so detected, they can be corrected. Generally, for an n-bit code having p additional check bits, a single error can be corrected if

$$n \leqslant 2^p - p - 1$$

Thus, in the case of a four-bit data code, at least three additional check bits are required in order to detect and correct one error.

Perhaps the best known single-error correcting code is the Hamming code[1] shown in Table 1.5 (this code can also detect more than one error). This code uses even parity, each parity bit (P_1, P_2 or P_3) is taken in association with three data bits (the data code is the 8421 BCD code). These groups are

$$P_1 841, P_2 821, P_3 421$$

The receiving equipment generates three even parity bits P_1', P_2' and P_3' corresponding to the three groups listed above. If the check combination

Table 1.5

A Hamming single-error correcting code

Number	P_1	P_2	8	P_3	4	2	1
0	0	0	0	0	0	0	0
1	1	1	0	1	0	0	1
2	0	1	0	1	0	1	0
3	1	0	0	0	0	1	1
4	1	0	0	1	1	0	0
5	0	1	0	0	1	0	1
6	1	1	0	0	1	1	0
7	0	0	0	1	1	1	1
8	1	1	1	0	0	0	0
9	0	0	1	1	0	0	1

$P'_3P'_2P'_1$ is zero, the incoming data is assumed to be correct; if it has a finite value, then the bit at the position in the code indicated by the decimal value $P'_3P'_2P'_1$ (commencing at the left or m.s.b. of the incoming code group) is in error. Consider the example below. Suppose that the binary equivalent of 6_{10} (1100110_2 in the Hamming code) is transmitted but, due to a transmission error, the code group 1100100_2 is received. Applying an even parity check to $P_1 841$ (1010) yields $P'_1 = 0$, the check applied to $P_2 821$ (1000) yields $P'_2 = 1$, and to $P_3 421$ (0100) yields $P'_3 = 1$. The resultant check word $P'_3P'_2P'_1 = 110_2 = 6_{10}$, indicates an error in the sixth bit of the code group (commencing with P_1).

According to Hamming, the best code construction is obtained when the p parity bits are placed in positions 1, 2, 4, 8, ... 2^{p-1} from the left or m.s.b. of the code word. The additional parity bits are sometimes known as the *Hamming supplement*.

1.10 BCD codes with more than four bits

Although only four bits are needed to represent all ten decimal digits, codes using more than four bits provide simple means of checking for errors; in general, these codes require very simple binary-to-decimal decoders. A disadvantage of some of these codes is that arithmetic processes are difficult with them. Codes with weightings of 5—0—4—3—2—1—0, 7—4—2—1—0, 5—4—3—2—1—0 and 5—1—1—1—1 have been used. 'Unweighted' codes can also be used. Three codes are listed in Table 1.6.

The biquinary code in Table 1.6 is one of a family known as *m-out-of-n* codes[2], in which each group of bits is represented by n bits, m of which have the value '1'. These codes are also described as $\binom{n}{m}$ codes. The biquinary code is a $\binom{7}{2}$

Table 1.6

Code identification	biquinary	7—4—2—1—0	Walking or creeping code
Code type	2-out-of-7	2-out-of-5	
Weight	5043210	74210	NO
0	0100001	11000	00000
1	0100010	00011	00001
2	0100100	00101	00011
3	0101000	00110	00111
4	0110000	01001	01111
5	1000001	01010	11111
6	1000010	01100	11110
7	1000100	10001	11100
8	1001000	10010	11000
9	1010000	10100	10000

code, in which each binary code group contains two 1's and five zeros; in this case, the left-hand '1' is called the binary bit, and the right-hand '1' the quinary bit. Quibinary codes are used, one combination being 8—6—4—2—0—1—0.

The 7—4—2—1—0 is a $\binom{5}{2}$ code. Decimal numbers are represented in this code. by equivalent weights in the 2-out-of-5 code, zero being represented by a number which is in excess of the normal decimal count, i.e., by a value greater than 9_{10}; the only possible combination is 11000.

The *walking code* or *creeping code* in Table 1.6 gives the appearance of the 1's creeping from the right to the left, followed by zeros creeping through the number. This code is a *unit-distance code*, in which the maximum distance between any adjacent pair of groups is unity. This code can be generated by a simple five-stage twisted-ring counter (see section 7.7).

1.11   Excess-three binary-decimal code

This code is formed by adding the binary equivalent of decimal three to the pure binary number. It is referred to as the XS3 code, and is shown in Table 1.7. It has a number of advantages. Thus it is easy to calculate, it is simple to translate into pure binary, and no number is represented by four 0's. A feature of the

Table 1.7

The excess-three code

Decimal number	XS3 code
	8421
0	0011
1	0100
2	0101
3	0110
4	0111
5	1000
6	1001
7	1010
8	1011
9	1100

excess-three code is that it is self-complementing (see also section 1.7). An interesting property of the code is that if the two least significant bits (the '2' and the '1' bits) are complemented, the resulting code has weights of 8, 4, (−2), (−1).

1.12   Position sensing codes

When a pure binary coded plate or disc is used to encode the position of an object, ambiguity can arise since several of the bits change simultaneously. This

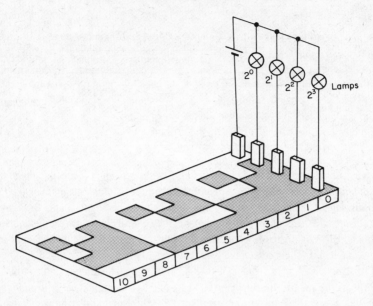

Fig. 1.1   Diagram of a plate, coded in pure binary, for use as a position sensing device.

is illustrated for a pure binary linear plate coded in Fig. 1.1 and Table 1.8. The figure shows a conducting plate with areas of insulating material, shaded, attached to its surface. Connection to the plate is made by five brushes, one being the common return connection for the electricity supply. The linear position of the plate is given by the combination of the lamps illuminated and extinguished.

Table 1.8

Pure binary position sensing code

Position	Pure binary code 8421	Number of simultaneous changes in binary digits
0	0000	
1	0001	1
2	0010	2
3	0011	1
4	0100	3
5	0101	1
6	0110	2
7	0111	1
8	1000	4
9	1001	1
10	1010	2

Table 1.9

A typical transition sequence for a pure binary coded plate

	Pure binary	Decimal
Initial position	0111	7
Possible outputs $\left\{\vphantom{\begin{matrix}a\\b\\c\end{matrix}}\right.$ during transition	1111	15
	1101	13
	1100	12
Final position	1000	8

When the change 7_{10} to 8_{10} occurs all four bits change. To obtain the change simultaneously in a mass-produced product is to ask almost for the unobtainable. The accuracy with which brushes can be aligned, and the accuracy of production of the surface treatment on the plate imposes severe restrictions on this method of position measurement. Optical methods using discs and plates are also used, but these also have limitations.

If, during the change from 7_{10} to 8_{10}, the 2^3 brush changes contact before the others, the maximum output of 1111_2 or 15_{10} is indicated. If it changes after the others, the minimum output of 0000 is indicated during the transition. A typical sequence of events is shown in Table 1.9.

In practice, position sensing devices usually employ codes which limit the number of digits changed during the transition to either two or one. The latter group are unit-distance codes.

1.12.1 Binary-decimal position sensing codes

Two examples are given in Table 1.10. Code 1 is a biquinary code; it uses seven digits to cover the decimal range 0 to 9, the maximum number of bits changed during any transition is two. This occurs during the change between decimal 4 and 5. A form of parity check exists, since each decimal number is represented by a binary combination of two 1's and five 0's. Code 2 is a unit-distance code requiring four digits.

Table 1.10

Examples of binary-decimal position sensing codes

Decimal	Code 1	Code 2
0	0100001	0000
1	0100010	0001
2	0100100	0011
3	0101000	0010
4	0110000	0110
5	1010000	1110
6	1001000	1010
7	1000100	1011
8	1000010	1001
9	1000001	1000

Arithmetic processes are difficult to carry out directly in both codes, and for this purpose they are converted into pure binary code. This is dealt with in detail in chapter 7.

1.12.2 Cyclic or reflected binary codes

The most useful group of unit-distance codes are the *reflected binary* codes. The *Gray code* is the most popular of these. It is built up of a simple 0 and 1 combination which is progressively reflected as the system builds up in magnitude. The first two numbers in the Gray code are 0000 and 0001, representing zero and unity, respectively. The next two numbers are obtained by reflecting the combination, shown dotted below, and adding a '1' in the next higher digit.

Decimal	Gray code
0	0 0 0 0
1	0 0 0 1
2	0 0 1 1
3	0 0 1 0

The first fifteen numbers of the Gray code are given in Table 1.11 to illustrate the build up of the system. Arithmetic processes are difficult to carry out in the Gray code, and it is usual to convert it to pure binary before carrying out mathematical operations on the data. Unit-distance codes are discussed in more detail in section 3.9.

Table 1.11

The Gray code

Decimal	Gray code
0	0000
1	0001
2	0011
3	0010
4	0110
5	0111
6	0101
7	0100
8	1100
9	1101
10	1111
11	1110
12	1010
13	1011
14	1001
15	1000

Reflected BCD or XS3 codes can be derived by using groups of ten of the Gray code combinations, as shown in Table 1.12. A feature of the reflected XS3 code is that the 9's complement (see section 2.2.1) of any number is obtained by complementing the most significant digit.

Table 1.12

Reflected BCD and XS3 codes

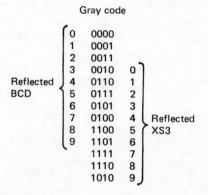

1.13   Alphanumeric codes

Many computers and systems handle both alphabetical and numerical information. One six-bit code dealing with both forms of information is shown in Table 1.13. The two left-hand bits serve to distinguish between numerical and alphabetical information, an 8421 BCD weighting being used for the numerical section of the code.

Table 1.13

An alphanumeric code

Decimal	Code	Letter	Code	Letter	Code	Letter	Code
0	00 0000	A	01 0000	K	10 0000	U	11 0000
1	00 0001	B	01 0001	L	10 0001	V	11 0001
2	00 0010	C	01 0010	M	10 0010	W	11 0010
3	00 0011	D	01 0011	N	10 0011	X	11 0011
4	00 0100	E	01 0100	O	10 0100	Y	11 0100
5	00 0101	F	01 0101	P	10 0101	Z	11 0101
6	00 0110	G	01 0110	Q	10 0110		
7	00 0111	H	01 0111	R	10 0111		
8	00 1000	I	01 1000	S	10 1000		
9	00 1001	J	01 1001	T	10 1001		

In order to transmit a complete text in the English language, even if only capital letters are transmitted, at least 52 characters are required as follows

    26 code words for the letters A–Z
    10 code words for punctuation marks
    10 code words for the digits 0–9
     6 code words for typewriter instructions.

The latter six instructions are required for the purpose of causing the electrical typewriter to function correctly, and include such instructions as carriage return, line feed, space, etc. A 6-bit code can represent 64 characters, leaving 12 characters for special symbols.

A range of alphanumeric codes has been developed for the purpose of text transmission, a popular code being the ASCII teleprinter code (American Standard Code for Information Interchange). The code uses an 8-bit word, seven of them being used to code a total of 128 characters, the 8th bit providing an even parity check.

Problems

1.1. Discuss the advantages and disadvantages of the pure binary code, when compared with decimal, as applied to electronic computing circuits.

1.2. Translate the following decimal numbers into their pure binary equivalents.
(a) 5268,   (b) 23·75,   (c) 0·0125

1.3. Convert the following into their decimal equivalents,
(a) $(7852)_9$,   (b) $(2734)_8$,   (c) $(253)_6$,   (d) $(421)_5$

1.4. Convert the following fractions into pure binary numbers.
(a) $\frac{1}{16}$,   (b) $\frac{7}{8}$,   (c) $\frac{2}{3}$,   (d) $\frac{17}{32}$

1.5. Convert the following pure binary numbers into decimal.
(a) 011101010111011,   (b) 0·0111,   (c) 10111·01

1.6. Convert the following decimal numbers into numbers with the radix indicated.
(a) 989 into radix 8,   (b) 732 into radix 6
(c) 876 into radix 5,   (d) 932 into radix 3

1.7. Convert the following numbers into the numbering systems indicated.
(a) $(857)_9$ into radix 6,   (b) $(231)_4$ into the Gray code
(c) $(576)_8$ into radix 9,   (d) $(222)_5$ into radix 3

1.8. Write down a 2421 BCD code not given in this chapter.

1.9. Write down two different 3321 decimal code sequences.

1.10. Write down two different unit-distance decimal codes.

1.11. Convert 5249_{10} into (a) 8421 BCD, and (b) 5211 BCD.

1.12. Convert the Gray code combination 11101 into decimal.

1.13. Explain the terms: *bit, parity, binary-coded-decimal*. In the following group of binary words, four contain data, and one is a parity check word. The right-hand digit of each data word is also a parity check on that word.

1st word	*2nd word*	*3rd word*	*4th word*	*Parity check*
10011	01100	01011	01110	01101

The data words are coded in 8421 BCD. Determine the parity system used, the bit which is in error, and the corrected decimal number.

References

1. HAMMING, R. W., 'Error Detecting and Error Correcting Codes', *Bell System Technical Journal,* **29**, 147–160, 1950
2. GILMOUR, W. D., 'P-out-of-Q Codes', *Electron. Engng,* **35**, 41–43, 1963

2. Arithmetic processes

2.1 Addition

Where $x + y = z$, x is the *addend*, y the *augend*, and z the *sum*. In any numbering system of radix r, the maximum number that appears in any column is $(r - 1)$. If the sum exceeds this value, a *carry* is generated which must be added to the next higher digit. The carry digit is said to be *carried in* to a sum if it is the carry digit generated by the next lower significant digit. It is said to be *carried out* if the carry digit is generated by the addition process under consideration. This is illustrated below.

	Duodecimal	Decimal	Octal	Binary
Addend	9 1 6	9 1 6	7 7 7	1 0 1 0
Augend	+1 7 6	+1 7 6	+6 5 5	+1 0 1 1
Carry in	0 1 0	1 0 1 0	1 1 1 0	1 0 1 0 0
Carry out	0 0 1	0 1 0 1	0 1 1 1	0 1 0 1 0
Sum	t 9 0	1 0 9 2	1 6 5 4	1 0 1 0 1

A binary sum exists if either the addend or augend is present, and a carry is generated if both are present. This property allows networks to be designed which can provide the outputs compatible with the arithmetic process of summation.

2.1.1 Addition in binary-decimal code

Providing that the sum of the two numbers in 8421 BCD is less than 9_{10}, the rules for normal binary addition are followed, as shown below.

Decimal	8421 BCD	Decimal	8421 BCD
2	0010	7	0111
+3	+0011	+2	+0010
5	0101	9	1001

If the sum is greater than nine, one of the forbidden codes, e.g., the pure binary equivalents of 10_{10} to 15_{10}, is generated. This is illustrated below.

Decimal	8421 BCD
4	0100
+8	+1000
12	1100 (uncorrected sum)

The correct 8421 BCD sum should have been 0001, 0010. To obtain the correct solution two approaches are possible.

(a) Add 6_{10} (0110_2) to the result.
(b) Subtract 10_{10} (1010_2) from the result and add a '1' in the 10's column of the 8421 BCD sum.

This is illustrated as follows.

	Method (a)	Method (b)
Uncorrected sum	1100	1100
	+0110	−1010
		0010
		+0001, 0000
Sum	0001, 0010	0001, 0010

In method (a) the carry digit is generated automatically when 6_{10} is added, whereas in method (b) it must be generated by other means.

When the sum of the BCD digits is greater than 15_{10}, the binary sum generates a 'carry', but the sum is not correct, as shown below.

Decimal	8421 BCD
8	1000
+9	+1001
17	0001,0001 (uncorrected sum)

The corrected sum is obtained by adding 6_{10} (0110_2) to the result. The simple rule for adding 8421 BCD may therefore be stated as follows.

If the binary sum of the numbers generates a forbidden code, or a 'carry', add the 8421 BCD code for 6_{10} to the resulting sum, otherwise the result is correct.

2.1.2   Addition in excess-three code

The XS3 code is given in Table 1.7. A feature of the code is that no decimal digit is represented by four zeros. The 9's complement (see section 2.2.2) is obtained by converting the 0's into 1's and the 1's into 0's. This process is known as *complementing* the digits.

When the sum of two XS3 numbers is 9_{10} or less, the uncorrected sum in the XS3 code is 3_{10} greater than it should be. This is illustrated below.

Decimal	XS3
3	0110
+3	+0110
6	1100 (uncorrected sum)

The uncorrected XS3 result is equivalent to 9_{10}, which is 3_{10} in excess of the correct value. To compensate for this, 3_{10} is subtracted from the uncorrected sum. When the sum of the two numbers is greater than 9_{10}, the uncorrected XS3 sum is 3_{10} less than it ought to be. To correct the XS3 result, 3_{10} is added to the result as shown below.

Decimal	XS3	
8	1011	
+9	+1100	
	0001,0111	(uncorrected sum)
	+0011	(3_{10})
17	..., 1010	(corrected sum).

The simple rule for XS3 addition is to add 0011_2 when a 'carry' is generated, otherwise subtract 0011_2.

2.2 Subtraction

Where $x - y = z$, x is the *minuend*, y the *subtrahend*, and z the *difference*. Provided that x is greater than y, z is positive and the subtraction proceeds normally. When the subtrahend is greater than the minuend, a *borrow-in*, equal in value to the radix r from the next higher x digit, must be made. To keep the problem unchanged a *borrow-out* must be added to the next higher value of y. This is illustrated below for various systems.

	Duodecimal	Decimal	Octal	Binary
x	t 1 6	6 7 8	5 4 3	1 0 1 1
y	–9 7 e	–5 8 9	–4 7 7	–0 1 1 1
Borrow-in	0 10 10	0 10 10	0 10 10	0 10 0 0
	✓ ✓	✓ ✓	✓ ✓	✓
Borrow-out	1 1 0	1 1 0	1 1 0	1 0 0 0
z	0 5 7	0 8 9	0 4 4	0 1 0 0

2.2.1   Representation of negative numbers

A negative number may be represented either in the *signed modulus notation* or in a *complement notation*. Man is most familiar with the signed modulus notation, in which the number is written in the form

$$(\text{sign}) \, (\text{modulus})$$

Thus the number -8_{10} is a value which is eight digits below zero (or is less than zero), and -101_2 is 5_{10} bits below zero. The negative sign is a man-made philosophical concept which electronic systems cannot understand, and the negative sign must be converted into an acceptable form for handling by the electronic system. When numbers are stored in the signed modulus form, one of the digits (or bits) is used in lieu of the 'sign'. In the decimal system this digit is described as a *sign digit*, which is usually assigned the value zero in the case of a positive number, and is nine in the case of a negative number. In the binary system the additional bit is known as the *sign bit*, whose value is usually zero for positive numbers and is unity for negative numbers. Thus

$$-8_{10} \equiv (9)8_{10}$$
$$-101_2 \equiv (1)101_2$$
$$+110_2 \equiv (0)110_2$$

For the convenience of readers, the sign digit (the most significant digit) is enclosed in parenthesis; when stored in a machine, the sign digit is identified by the fact that it occupies the most significant position in the number.

2.2.2   Decimal complement notation

Every ordered numbering system has a radix complement or true complement. In the decimal system this is referred to as the *tens complement*. The 10's complement of a number A, of order m, is $(10^m - A)$.

Suppose that we wish to evaluate the 10's complement of the number 1086_{10}; readers are reminded that in electronic storage systems the most significant position in the number is reserved for the sign digit, and the number is effectively $+1086 \equiv (0)1086$. For this number $m = 5$, and its 10's complement is $10^5 - (0)1086 = (9)8914$. It is important to note that normal arithmetic procedures are applied to the sign bit; the sign bit (9) of the number so obtained above indicates that the number stored has a negative value, and that it is stored in the 10's complement notation. Thus

$$-1086_{10} \equiv (9)8914$$

Since the above number contains a sign digit, it is presented in the *signed complement notation*. A simple rule for determining the 10's complement of a decimal number is as follows: *subtract each digit (including the sign digit) from*

9 and, to the resulting difference, add '1' to the least significant digit. This rule is also applied to decimal numbers containing a fractional part.

A consequence of the above is that the storage capacity of the system should be at least one digit greater than the storage capacity required for the data. If a number of order m is to be stored, the system must be capable of storing m digits. To show how the signed complement notation functions in practice, the number 1086_{10} is subtracted from 2000_{10}. The difference between the two numbers is obtained by *adding* the signed complement of 1086 to 2000, as follows.

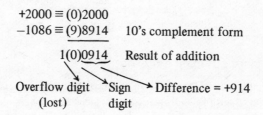

$$+2000 \equiv (0)2000$$
$$-1086 \equiv (9)8914 \quad \text{10's complement form}$$

$$1(0)0914 \quad \text{Result of addition}$$

Overflow digit   Sign   Difference = +914
(lost)   digit

The result above yields *a sign digit of zero, which indicates that the result is the true difference which has a positive value*. It also produces an overflow digit which, since it is in excess of the storage capacity of the system, is lost.

Should the calculation produce a sign bit of (9), then the resulting difference has a negative value, and the result is the 10's complement of the true difference. The true difference can be computed by the method outlined for determining the 10's complement, described earlier.

Yet another form of complement notation in common usage is known as the *diminished radix complement*, and for a number A of order m is

$$(r^m - 1) - A$$

where r is the radix of the system. Comparing this definition with that of the radix complement, we see that it has a value which is numerically less than the radix complement by a factor of unity. In the decimal system, the diminished radix complement is known as the *nines complement*. *The 9's complement of a decimal number is obtained by subtracting each digit (including the sign digit) from nine.* For example, the 9's complement of $(0)1086$ is $(9)9999 - (0)1086 = (9)8913$. The subtraction process must be modified when compared with that of the radix complement, to account for the '1' difference between the two complement notations. This is done simply by adding the overflow digit (which may either be a '1' or a '0') to the least significant digit of the sum. The latter process is known as *end-around carry*, and is illustrated in the following example.

Example 2.1: Subtract 232_{10} from 1036_{10} using the 9's complement notation.

Solution: The 'length' of both numbers must first be adjusted to be equal to one another, i.e., the numbers become (0)0232 and (0)1036, respectively.

$$9\text{'s complement of } (0)0232 = (9)9767$$

$$
\begin{aligned}
+1036 &\equiv (0)1036 \\
-232 &\equiv (9)9767 \\
\hline
1(0)0803 \quad & \text{uncorrected sum}
\end{aligned}
$$

$\longrightarrow 1$ 'end-around carry'

(0)0804

Sign bit   Difference = +804

2.2.3   Binary complement notation

The *radix complement* (also known as the *true complement* or *twos complement*) of a number A, of order m (including the sign bit) is $(2^m - A)$. The *diminished radix complement* or *ones complement* is less than the radix complement by a factor of unity, and is given by $[(2^m - 1) - A]$. Simple rules for the derivation of these complements are given below.

2's complement:
(a) Change the 0's to 1's and 1's to 0's, and add '1' to the least significant bit.
(b) Copy the number commencing with the least significant bit up to and including the least significant '1', all bits thereafter being complemented or inverted.

1's complement:
Change the 0's into 1's and 1's into 0's.

When the number is stored in an electronic system, a sign bit is assigned to the number, and this is dealt with as though it were part of the modulus. Using rule (b) above, the 2's complement of (0)10·010 is obtained as follows

2's complement of $(0)10 \cdot 010 \equiv (1)01 \cdot 110$

inverted   unchanged

and the 1's complement of the same value is

1's complement of $(0)10 \cdot 010 \equiv (1)01 \cdot 101$

Further examples illustrating the above rules are given below

Binary number	2's complement	1's complement
(0)001·0	(1)111·0	(1)110·1
(0)100	(1)100	(1)011
(0)101·10	(1)010·10	(1)010·01
(0)000·01	(1)111·11	(1)111·10
(1)010·11	(0)101·01	(0)101·ōō

Arithmetic processes with binary complements are similar to those with decimal complements.

SUBTRACTION USING THE SIGNED 2's COMPLEMENT NOTATION

The general rule is that *the 2's complement of the subtrahend is added to the minuend. If the sign bit of the result is '0', then the result is the true difference, and is assigned a positive sign. If the sign bit is '1', the result is the 2's complement of the difference. Any overflow produced by the calculation is 'lost'.* Examples of the use of this notation are given below.

(a) $(6 \cdot 25 - 4 \cdot 25)_{10}$ or $((0)110 \cdot 01 - (0)100 \cdot 01)_2$

$$+6 \cdot 25_{10} = (0)110 \cdot 01_2$$
$$-4 \cdot 25_{10} = \underline{(1)011 \cdot 11} \quad \text{2's complement of } (0)100 \cdot 01$$
$$1(0)010 \cdot 00$$

lost

$$\text{Difference} = (0)010 \cdot 00_2 = +2_{10}$$

(b) $(-5 - 7)_{10}$

In all systems, care should be taken to ensure that the storage capacity is large enough to accommodate any intermediate or final result. In this case it must be large enough to deal with a total of 12_{10}; the binary storage capacity must be at least five bits (four for data and one for the sign bit). Let us assume that in this case the binary word has a length of seven bits, six of them being used for data.

$$+5_{10} = (0)000101_2$$

whose 2's complement is $(1)111011$, and

$$+7_{10} = (0)000111$$

whose 2's complement is $(1)111001$

$$-5_{10} = (1)111011 \quad \text{2's complement form}$$
$$-7_{10} = \underline{(1)111001} \quad \text{2's complement form}$$
$$1(1)110100$$

lost

Result = 2's complement of $(1)110100$
$$= -(0)001100_2 = -12_{10}$$

SUBTRACTION USING THE SIGNED 1's COMPLEMENT NOTATION

In this case, *the 1's complement of the subtrahend is added to the minuend, and the overflow bit (not the sign bit) is added to the least significant bit of the uncorrected sum. If the sign bit of the corrected sum is '0', then the result is the true difference. If the sign bit is '1', then the result is the 1's complement of the difference.* Examples of the use of this notation are given below.

(a) $(9 \cdot 5 - 5 \cdot 25)_{10}$ or $((0)1001 \cdot 10 - (0)0101 \cdot 01)_2$

$$+9 \cdot 5_{10} \ = (0)1001 \cdot 10_2$$
$$-5 \cdot 25_{10} = \underline{(1)1010 \cdot 10_2} \qquad \text{1's complement of } (0)0101 \cdot 01$$

$$1(0)0100 \cdot 00 \qquad \text{uncorrected sum}$$

$$\longrightarrow 1 \qquad \text{'end-around carry'}$$

$$(0)0100 \cdot 01 \qquad \text{corrected sum}$$
$$\text{Difference} = (0)0100 \cdot 01_2 = +4 \cdot 25_{10}$$

(b) $(4 \cdot 25 - 6 \cdot 25)_{10}$ or $((0)100 \cdot 01 - (0)110 \cdot 01)_2$

$$+4 \cdot 25_{10} = (0)100 \cdot 01_2$$
$$-6 \cdot 25_{10} = \underline{(1)001 \cdot 10_2} \qquad \text{1's complement of } (0)110 \cdot 01$$

$$0(1)101 \cdot 11 \qquad \text{uncorrected sum}$$

$$\longrightarrow 0 \qquad \text{'end-around carry'}$$

$$(1)101 \cdot 11 \qquad \text{corrected sum}$$
$$\text{Difference} = (1)101 \cdot 11_2 = -(0)010 \cdot 00_2 = -2_{10}$$

2.3 Multiplication

Multiplication is a form of repeated addition. The calculation 271 x 23 can be performed in several ways, viz.:

(a) Add 271 to 0 twenty-three times.
(b) Add 271 to itself twenty-two times.
(c) Add 271 to 0 twice, shift one place to the left and add 271 three times to the sum.

The rules for multiplication of binary numbers are, fortunately, simpler than those for other systems. The product of two variables, x and y, is always zero except in the case when both x and y are unity, when the product is '1'. An illustrative example is given below.

Decimal	Binary	
9	1001	
9	1001	
	1001	
	0000	
	0000	
	1001	
81	1010001	Product

The rule for multiplication by the above method is to add when the multiplier is 1, and shift one place to the left. When the multiplier is 0, the result is simply shifted two places to the left.

The process of shifting and adding is mechanised in digital systems by means of electronic stores known as *registers* (see chapter 7 for details). If two numbers A and B are to be multiplied together, the multiplicand (A) is stored in one register, and the multiplier (B) is stored in a second register. The product is stored in a register (P) whose 'length' is double that of the data words. Register A retains the same data throughout the operation, whilst the contents of register B are shifted to the right after each stage of the calculation. When multiplying two binary numbers, each of length N bits, N add and shift steps are carried out. A typical example is illustrated below. Here $A = 110_2$ and $B = 101_2$; the least significant bit of number B is known as the *multiplication bit, M*. The sequence begins with an addition and is terminated with a shift-right, the processes being:

Addition: If $M = 0$, add zero to the most significant half, P_M (see Table 2.1), of register P. If $M = 1$, add the contents of register A to P_M.

Shifting: Shift the contents of register B and the contents of register P one place to the right.

The complete process for the numbers given above is illustrated in Table 2.1.

If either of the two numbers has a negative sign, one of the two following methods can be used to evaluate the product. In one method the sign bits are compared and, if these are not equivalent to one another, then the product of

Table 2.1

Contents of register A	Contents of register B	Contents of the double-length product register, P	
110	101 ↑	000 000	
multiplication bit, M		P_M	
			$M = 1$, add A to P_M
		110 000	
			Shift B and P one place to the right
110	10	011 000	
			$M = 0$, add zero to P_M
		011 000	
			Shift B and P one place to the right
110	1	001 100	
			$M = 1$, add A to P_M
		111 100	
			Shift B and P one place to the right
110		011 110	
			Stop
		Product = 011110_2	

the moduli has a negative sign. In the second method the numbers are multiplied together directly and, if either has a negative sign, special algorithms are used to deduce the correct value of the product.

2.4 Binary division

Division can be carried out by a process of repeated subtraction and shifting to the right. Consider $1010_2 \div 100_2$:

$$
\begin{array}{r}
10 \cdot 1 \\
100\ \overline{)\ 1010} \\
\underline{100} \\
10 \\
\underline{00} \\
100 \\
\underline{100} \\
000
\end{array}
$$

The process shown above can be mechanised, and is used in some digital systems and is known as the *restoring method* of binary division. An algorithm for restoring binary division is given below.

> Subtract the N bits of the *divisor* from the N most significant bits of the *dividend*. If the difference is *zero or positive*, enter a '1' into the least significant end of the partial quotient register, and leave the difference in the dividend register. Shift the contents of the dividend and partial quotient registers one place to the left. Repeat the subtraction process.
>
> If the difference is *negative*, enter a '0' in the least significant end of the partial quotient register. Add the divisor to the difference to *restore* the dividend to its former value. Shift the contents of the dividend and partial quotient registers one place to the left. Repeat the subtraction process.

The above process is illustrated using a divisor of 100_2 and a dividend of 1010_2 (see Table 2.2). The subtraction process is carried out using the signed 2's complement addition method; using this notation, $-100_2 \equiv (1)100$. Readers will note that the sign bit of the divisor is always aligned with the sign bit of the dividend.

The position of the binary point in the solution is calculated from the order of the dividend and of the divisor and, if these are p and q respectively, the power of the most significant bit of the quotient is 2^{p-q}. In the case of the example in Table 2.2, this is $2^{3-2} = 2^1$, hence the answer is 10.1_2.

The restoring method of division is a relatively slow operation due to the time-wasting process of restoring when a negative remainder is obtained. A faster process known as *non-restoring division* is used in most digital systems, in which

Table 2.2

Divisor	Dividend	Partial quotient	
(0)100	(0)1010		
	(1)100		Subtract divisor by complement addition
lost ←	1(0)0010	1	Remainder positive, record '1'
	(0)0100	1	Shift left
	(1)100		Subtract divisor
	(1)1100	10	Remainder negative, record '0'
	(0)100		Add divisor
lost ←	1(0)0100		Restored remainder
	(0)1000	10	Shift left
	(1)100		Subtract divisor
lost ←	1(0)0000	101	Remainder zero, record '1'

an artifice is used to eliminate the unwanted steps. The process is not fully described here, but an algorithm is given below.

Subtract the divisor from the dividend. If the difference is *zero or positive*, record a '1' in the partial quotient register. Shift the difference and the partial quotient one place to the left. Then *subtract* the divisor from the difference.

If the difference is *negative*, record a '0' in the partial quotient register. Shift the difference and the partial quotient to the left. Then *add* the divisor to the difference.

2.5 Floating point arithmetic

If calculations are carried out using normal radix point notation, the electronic system has the chore of keeping track of the radix point. This problem is overcome in the *floating point* representation of numbers, since the position of the radix point is explicitly fixed at a point in the stored number. A floating point number, X, is represented in the form

$$X = m\, r^e$$

where m is known as the *mantissa* (or as the *fractional part* or *argument*) of the number, r is the radix of the system, and e is the *exponent* or *characteristic* of the number. In electronic machines, the data and results are scaled so that the most significant data bit is a '1'. The radix point lies between this bit and the sign bit. A number is said to be *normalized* when presented in this form. The following are examples of normalized floating point representation.

$$1011_2 = +{\cdot}1011 \times 2^4 \tag{1}$$
$$1{\cdot}001_2 = +{\cdot}1001 \times 2^1 \tag{2}$$
$$-10{\cdot}1_2 = -{\cdot}101 \times 2^2 \tag{3}$$
$$0{\cdot}0101_2 = +{\cdot}101 \times 2^{-1} \tag{4}$$

In the cases of (1) and (2) above, both the mantissa and the exponent would be stored in normal binary notation. In the case of (3), the mantissa would be stored in a binary complement notation, as would the exponent in (4) above.

After an arithmetic operation the result may not be in its normalized form; a process known as *standardizing* is carried out on the number to return it to its normalized form. The principal mathematical processes are summarized below.

FLOATING POINT ADDITION AND SUBTRACTION
The operands are scaled so that their exponent values are equal. To do this, the smaller value is shifted right until the exponents are equal in value.

$$(0{\cdot}111 \times 2^2) + (0{\cdot}1 \times 2^0) = (0{\cdot}111 \times 2^2) + (0{\cdot}001 \times 2^2)$$
$$= 1{\cdot}000 \times 2^2$$

The solution is standardized to give the solution $0{\cdot}1 \times 2^3$.

FLOATING POINT MULTIPLICATION AND DIVISION
When multiplying, the mantissas of the operands are multiplied together, and the exponents are added. To divide, the mantissa of the dividend is divided by the mantissa of the divisor, and the exponent of the divisor is subtracted from that of the dividend. These are expressed as follows.

$$Ar^a \times Br^b = AB\, r^{(a+b)}$$
$$Ar^a \div Br^b = \frac{A}{B}\, r^{(a-b)}$$

FLOATING POINT NUMBER REPRESENTATION
Floating point numbers are generally stored in digital systems in the general format shown in Fig. 2.1, the mantissa and the exponent having specific storage locations. In the case considered the word is 32 bits long, eight being allocated to the exponent and 24 to the mantissa. A combination of eight bits is frequently referred to as a *byte*.

In some cases the exponent is stored in what is known as a *biased binary notation* or *offset binary code*. In this notation, positive exponents are stored in

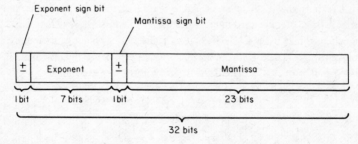

Fig. 2.1 A method of storing a floating point number.

the natural binary code together with a complemented most significant bit (the sign bit). This is equivalent to using binary '1' to represent the sign bit of positive numbers. Negative numbers are stored in the 2's complement notation, the m.s.b. again being complemented. This has the effect of 'biasing' the code so that the largest negative exponent is represented by a group of zeros, and the largest positive value by a group of 1's. A 5-bit code group can be used to store exponents in the range −16 to +15 using the biased binary notation, examples of which are given in Table 2.3.

Table 2.3

Examples of an offset binary code	
True value of exponent	Biased binary representation
15	11111
14	11110
.	.
.	.
1	10001
0	10000
−1	01111
.	.
.	.
−15	00001
−16	00000

Problems

2.1. Convert the following decimal numbers into pure binary numbers, and add them together, using binary arithmetic.
(a) $4 + 7$,   (b) $16 + 20$,   (c) $17 \cdot 5 + 12 \cdot 75$
(d) $5\frac{5}{16} + 2\frac{3}{8}$,   (e) $7 \cdot 5 + 5 \cdot 3 + 5\frac{7}{8}$

2.2. Convert the following decimal numbers into pure binary numbers, and subtract them using binary arithmetic.
(a) $5 - 2$,   (b) $20 - 10$,   (c) $4 - 5$
(d) $6 \cdot 25 - 4 \cdot 25$,   (e) $7 - 2 \cdot 5 - 3\frac{3}{4}$

2.3. Multiply the following numbers together, using the binary notation.
(a) 6×3,   (b) $7 \cdot 5 \times 3 \cdot 75$,   (c) $0 \cdot 25 \times 4 \cdot 25$
(d) $0 \cdot 5 \times 4 \times 0 \cdot 3$,   (e) -2×6

2.4. Using only binary arithmetic, compute the solutions to the following examples in which the values are given in denary. Give the solutions in binary.
(a) $6 \div 3$,   (b) $7 \cdot 5 \div 2 \cdot 5$,   (c) $1 \cdot 125 \div 6$,
(d) $17 \cdot 5 \div 0 \cdot 5 \div 3 \cdot 5$,   (e) $24 \cdot 375 \div 3\frac{3}{4}$

3. Fundamentals of logic

It is ironical to reflect that the foundation of modern logic theory has lain
dormant for many years before being utilized in practice. The basis of logic as
we now know it was set down in the nineteenth century by Boole, De Morgan,
and others. Many generations were to pass before technologists began to realize
the implications of these theories, and even longer for the fabrication techniques
for the construction of 'logical' devices to be perfected.

3.1   Traditional logic

Traditional logic can be illustrated by considering groups of people, shown in
Fig. 3.1. The whole population of any town, country, or continent can be split
up into many distinct groups: three are shown in the figure. The selection of the
groups is quite arbitrary, but every person falls into one or more of the
categories shown. If he does not, he does not exist.

Thus a man is either a sailor (S) or a non-sailor ($\bar{S}$). The bar over the S
signifies *negation*, or *logical inversion*, and is referred to as the logical NOT
function. That is,

$$\bar{S} = \text{NOT } S$$

A man who is married AND who is a sailor falls into the category $\bar{M}. S$. The
'dot' (.) is referred to as the logical AND function, or *connection*. This dot is
frequently confused with the dot used to represent the arithmetic product
function, but throughout this book it is used as the logical AND function. To
make matters worse the AND operation is often described as the *logical product
function*. Thus

$$\bar{M}. S = \text{a man who is unmarried AND is a sailor.}$$

Whenever equality signs (=) are used in logical equations they refer to logical
equality, which may not have the same meaning as arithmetic equality.

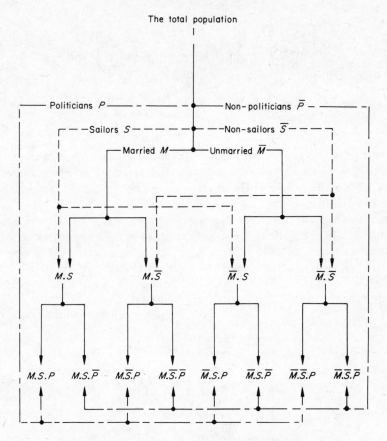

Fig. 3.1   Arbitrary divisions of the population.

For each group of people there are two possible divisions. Thus the whole population can be split into married people (M), and unmarried people ($\bar{M}$). Alternatively, the population can be split into sailors (S) and non-sailors ($\bar{S}$), or into politicians (P) and non-politicians ($\bar{P}$). When the intersections of any two groups are considered there are 2^2 combinations. For instance, between groups M and S these combinations are $M.S, M.\bar{S}, \bar{M}.S$, and $\bar{M}.\bar{S}$. The number of intersections double each time an additional group is introduced. With three groups it is 2^3, shown in Fig. 3.1, and for n groups it is 2^n. In all cases the grand total of all the intersections cannot exceed the total population, which is therefore referred to as the whole or unity (1).

3.2   The Venn diagram

The condition for one division of the population, discussed in section 3.1, is shown diagrammatically in the Venn diagram in Fig. 3.2(a). The whole popu-

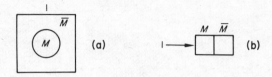

Fig. 3.2   Division of the Population, into married and unmarried people, by (a) the Venn diagram and (b) a rectangular form of representation.

lation is represented by the square, while those who are married are within the circle. Unmarried people are represented on the diagram by the area outside the circle, but within the square. The diagram can also be shown in rectangular form, as in Fig. 3.2(b).

All possible intersections of two groups are shown in Fig. 3.3. All who are married lie within the left-hand circle, while all sailors are encompassed by the right-hand circle. Thus circle M contains both married non-sailors ($M . \bar{S}$) and married sailors ($M . S$). The S circle contains married sailors ($M . S$) and unmarried sailors ($\bar{M} . S$). The area outside the circles is all that which is NOT within the two circles, i.e., unmarried non-sailors ($\bar{M} . \bar{S}$).

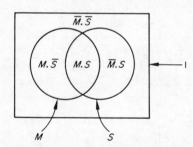

Fig. 3.3   Intersections of two groups of people on the Venn diagram.

The union of divisions of the population can also be investigated by the Venn diagram. Thus if we are interested in people who are married OR who are sailors, we find them within the shaded area in Fig. 3.4(a). People in the shaded area fall into the categroy ($M + S$). The 'plus' is referred to as the logical OR function, and should not be confused with the arithmetical addition sign. The OR function is sometimes referred to as the *logical sum function*. Thus in Fig. 3.4(a) the total function is

$$f_1 = \text{married people OR sailors} = M + S$$

Since the shaded parts of the diagram cover the areas represented by $M . \bar{S}$, $M . S$ and $\bar{M} . S$, it is clear that

$$f_1 = M + S = M . \bar{S} + M . S + \bar{M} . S$$

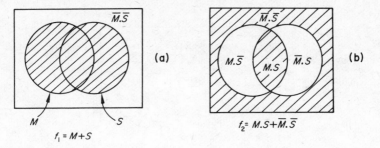

Fig. 3.4 (a) and (b)   The union of sections of the population.

In this way it is possible to construct logical equations, and to formulate logical proofs using the Venn diagram.

It is obvious from Fig. 3.4(a) that, since we are interested in people within the circles, we are NOT interested in the people outside the circles but who are within the square, i.e., NOT $\bar{M} \cdot \bar{S}$. Hence

$$M + S = \overline{\bar{M} \cdot \bar{S}}$$

In Fig. 3.4(b) the shaded areas cover people who are unmarried non-sailors $(\bar{M} \cdot \bar{S})$ OR who are married sailors $(M \cdot S)$, hence

$$f_2 = \bar{M} \cdot \bar{S} + M \cdot S$$

The unshaded area in Fig. 3.4(b) is NOT f_2, hence

$$\bar{f}_2 = \bar{M} \cdot S + M \cdot \bar{S}$$

From the expressions for f_2 and $\bar{f}_2$ it follows that

$$f_2 = \bar{M} \cdot \bar{S} + M \cdot S = \overline{\bar{M} \cdot S + M \cdot \bar{S}}$$

3.3   The Karnaugh map

As the number of groups of people considered increases, the Venn diagram becomes progressively more complex. Figure 3.5 illustrates the effect on the diagram of considering politicians (P) in addition to married people and sailors. In all there are eight areas on the diagram, corresponding to the eight results obtained in Fig. 3.1.

By allocating numbers in binary order to each group of people, each area on the diagram can be identified uniquely. By giving M the value 1_{10} (001_2), S the value 2_{10} (010_2), and P the value 4_{10} (100_2), the area defined as $P.S.M.$ has the value 7_{10} (111_2). Similarly $\bar{P}, \bar{S}$, and $\bar{M}$ are all given the value zero, and the area $\bar{P} \cdot \bar{S} \cdot \bar{M}$ has the value zero while $P \cdot S \cdot \bar{M}$ has the value 6_{10} (110_2).

Some simplification is effected by arranging the groups in a rectangular form, as shown in Fig. 3.6. All cells in the lower row are M, while cells in the upper row are $\bar{M}$. Similarly all cells in the second and third columns are P, while those

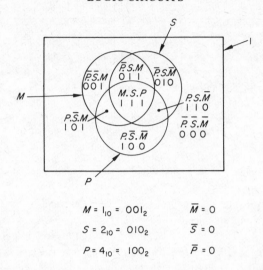

$$M = 1_{10} = 001_2 \qquad \overline{M} = 0$$

$$S = 2_{10} = 010_2 \qquad \overline{S} = 0$$

$$P = 4_{10} = 100_2 \qquad \overline{P} = 0$$

Fig. 3.5   Venn diagram for three groups of people within the total population

in the first and fourth columns are $\overline{P}$. The first two columns are defined as $\overline{S}$, while the third and fourth columns are S. This allows all possible intersections of M, S, and P to be shown on the diagram. This form of diagram is known as a *Veitch*[1] or *Karnaugh*[2] map.

Individual cells are defined in Fig. 3.6 either in terms of the alphabetical characters along the edges of the map, or in terms of the binary groups of digits along the edges. Veitch maps use alphabetical characters to define the cells, whilst the cells on the Karnaugh map are defined in terms of the binary values of variables P, S and M. The Veitch version is particularly useful when dealing with logical expressions of the type in section 3.4, and the Karnaugh version is more useful when transferring data from truth tables (see chapter 4). The name

PS M	00	10	11	01	
0	$\overline{P}.\,\overline{S}.\,\overline{M}$ 0 0 0$_2$ 0$_{10}$	$P.\,\overline{S}.\,\overline{M}$ 1 0 0$_2$ 4$_{10}$	$P.\,S.\,\overline{M}$ 1 1 0$_2$ 6$_{10}$	$\overline{P}.\,S.\,\overline{M}$ 0 1 0$_2$ 2$_{10}$	
1	$\overline{P}.\,\overline{S}.\,M$ 0 0 1$_2$ 1$_{10}$	$P.\,\overline{S}.\,M$ 1 0 1$_2$ 5$_{10}$	$P.\,S.\,M$ 1 1 1$_2$ 7$_{10}$	$\overline{P}.\,S.\,M$ 0 1 1$_2$ 3$_{10}$	M

Fig. 3.6   Veitch and Karnaugh representation of Fig. 3.5

Karnaugh map is used throughout this book to describe both Veitch and Karnaugh maps.

Karnaugh maps for two and four variables are shown in Figs. 3.7(a) and (b) respectively. The fourth group C contains those who possess a car, giving sixteen possible intersections in all ranging from $C.P.S.M$ to $\bar{C}.\bar{P}.\bar{S}.\bar{M}$.

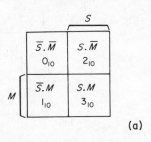

(a)

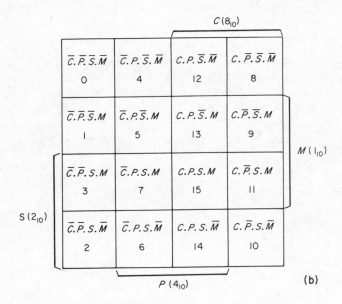

(b)

Fig. 3.7   Karnaugh map for (a) two and (b) four variables

3.4   Function mapping

Mapping is a graphical method of representing logical equations. It finds wide use in proving theorems, and in the design of logical networks. Consider the equations

$$f_1 = A.\bar{B}$$
$$f_2 = A.\bar{B}.\bar{C}.\bar{D}$$

The statement $f_1 = A \cdot \bar{B}$ implies that f_1 is a function of the two variables A and B, which exists ($f_1 = 1$) only when A AND (NOT B) occur simultaneously, otherwise the function does not exist ($f_1 = 0$). The function f_1 is therefore mapped as shown in Fig. 3.8(a). Function f_2 is mapped in Fig. 3.8(b).

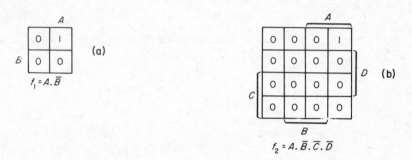

Fig. 3.8   Karnaugh maps for (a) $f_1 = A \cdot \bar{B}$ and (b) $f_2 = A \cdot \bar{B} \cdot \bar{C} \cdot \bar{D}$

It is also possible to obtain the logical product or sum of a number of Karnaugh maps, a common method in logical proofs. An example of the product of Karnaugh maps is given in Fig. 3.9. In Fig. 3.9 variables A and B are mapped independently, and the values of corresponding cells in each map are multiplied together to give the final function $f = A \cdot B$. For cell $\bar{A} \cdot B$ (lower left-hand cell), in the 'A' map, the value is '0', while in the 'B' map it is '1'. The value shown in the '$A \cdot B$' map is 0.1 = 0. Other cells are dealt with in this manner.

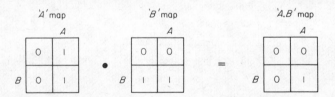

Fig. 3.9   Karnaugh map showing the logical product $A \cdot B$

An example of the logical sum of Karnaugh maps is given in Fig. 3.10. Provided that one of the two values in corresponding cells is zero, the logical sum follows normal arithmetic processes, thus

$$0 + 0 = 0$$
$$0 + 1 = 1$$
$$1 + 0 = 1$$

In both the 'A' and the 'B' maps, cell $\bar{A} \cdot B$ in Fig. 3.10 is marked with a '1'. It is at this point that the reader is reminded that the 'plus' sign is used here to

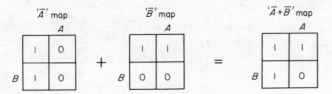

Fig. 3.10   The logical sum of two Karnaugh maps

represent the logical OR connective. Where this sum occurs, the following rule holds

$$1 + 1 = 1$$

A simple justification of this statement follows. Consider the logical equation $X + Y = Z$. Thus Z exists if X OR Y exists. By the same token Z also exists if both variables exist simultaneously.

Applying these rules sequentially to equivalent cells of the two left-hand maps in Fig. 3.10 gives the right-hand map.

3.5   Adjacent cells on the Karnaugh map

Cells which are adjacent on a Karnaugh map differ by only one binary digit (the weights of the digits being ignored), if the pure binary equivalent of the decimal notation in Fig. 3.7 is used. A four-variable map is shown in Fig. 3.11(a). Cell $A . B . \bar{C} . D$ contains three 1's, while those on either side have either two or four 1's. Cells at the top and bottom are adjacent in this sense, e.g., cells $\bar{A} . \bar{B} . \bar{C} . \bar{D}$ and $\bar{A} . \bar{B} . C . \bar{D}$.

Further inspection of the map shows that the four left-hand cells are adjacent to the four right-hand cells, according to the notation developed above, e.g., cells $\bar{A} . \bar{B} . \bar{C} . \bar{D}$ and $A . \bar{B} . \bar{C} . \bar{D}$.

For simplicity, the Karnaugh map is sometimes drawn in the manner shown in Fig. 3.11(b). The variables A and B have the binary combinations shown along the top, and variables C and D have the binary combinations shown on the left-hand side of the map. Thus, cell $A . B . \bar{C} . D$ has the binary value 1101, and is located at the point $AB = 11$, $CD = 01$. Cells $\bar{A} . \bar{B} . \bar{C} . \bar{D}$ (0000), $A . \bar{B} . \bar{C} . \bar{D}$ (1000), and $\bar{A} . \bar{B} . C . \bar{D}$ (0010) are located as shown.

Provided that cells are adjacent they may be grouped together. An example is shown in Fig. 3.12. Since each cell can be defined independently, the logical representation of the map is

$$f = A . B . C + A . B . \bar{C} + \bar{A} . B . \bar{C} + \bar{A} . B . C$$

Adjacent pairs of cells can now be grouped. In Fig. 3.12(a) pairs of cells in the horizontal plane are grouped, giving

$$f = B . C + B . \bar{C}$$

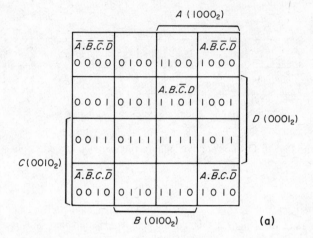

(a)

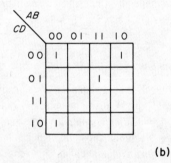

(b)

Fig. 3.11   The concept of adjacent cells is shown in (a). Map (b) shows an alternative way of displaying the cell code combinations

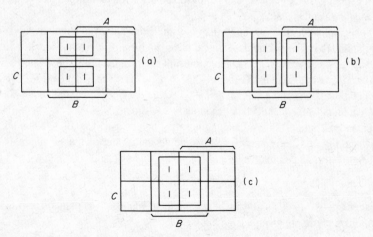

Fig. 3.12   Three methods of grouping cells are shown in (a), (b), and (c)

Grouping pairs of cells in the vertical plane, Fig. 3.12(b), gives

$$f = A . B + \bar{A} . B$$

Alternatively all four adjacent cells may be grouped as in Fig. 3.12(c) when

$$f = B$$

The number of cells grouped together must be in binary order, i.e., 1, 2, 4, 8, etc. In deducing logical equations from the Karnaugh map, it is essential that all the terms in the equation are marked by a '1', other cells are left blank or are marked with a '0'. If three adjacent cells appear on the map, then they are grouped as two pairs, illustrated in Fig. 3.13, where

$$f = A . B + B . C = B(A + C)$$

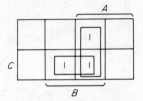

Fig. 3.13   Three adjacent cells are combined by grouping two pairs of adjacent cells

3.6   Don't care and can't happen conditions

It often happens in logical problems that certain combinations of the variables cannot be permitted to occur. Such a state is a *redundant* one, and is sometimes known as a *can't happen* condition. In some cases certain states are permitted to occur, but their existence does not affect the solution. This is a *don't care* condition, and is also redundant.

In section 2.1.1 it was shown that, in the 8421 BCD code, the combinations 1010, 1011, 1100, 1101, and 1111 are forbidden combinations of bits. If D represents the least significant digit (2^0), C represents the digit 2^1, etc., then the forbidden combinations are $A . \bar{B} . C . \bar{D}, A . \bar{B} . C . D, A . B . \bar{C} . \bar{D},$ $A . B . \bar{C} . D, A . B . C . \bar{D},$ and $A . B . C . D$ respectively. Each of these is plotted on the Karnaugh map in Fig. 3.14 by an X. Since these combinations 'can't

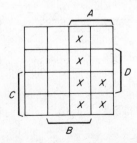

Fig. 3.14   The method of representing redundant states on the Karnaugh map

happen', it is feasible to include them in loops on the Karnaugh map without affecting the accuracy of the resulting equations.

In Fig. 3.15 the function defined is $f = A . B . \bar{C} . \bar{D} + A . B . C . \bar{D}$, together with 'can't happen' conditions $A . B . D$ and $\bar{A} . B . \bar{D}$. By including the redundant pair $A . B . D$, four cells are grouped as shown by the full line, reducing the equation to $f = A . B$. Alternatively using the redundant pair $\bar{A} . B . \bar{D}$, the cells grouped by the dotted lines define the expression $f = B . \bar{D}$. Both groupings are equivalent with the redundant conditions given.

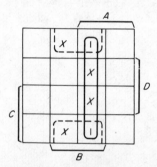

Fig. 3.15   Problem simplification can be achieved by utilizing redundant states

3.7   Complementing on the Karnaugh map

A function f is represented on the Karnaugh map by a series of 1's, and it is clear that $\bar{f}$ is represented by the cells filled with 0's. An example is shown in Fig. 3.16 for $f = A . B + A . C$. The complement of the function is obtained by grouping cells marked with zeros, in accordance with the rules laid down in this chapter. With the grouping shown in Fig. 3.16

$$\bar{f} = \bar{A} + \bar{B} . \bar{C}$$

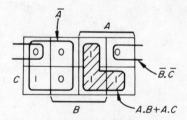

Fig. 3.16   Complementing by grouping the 0's

This technique can be used to define a group of 1's by saying that they are NOT the 0's. In Fig. 3.17, the four 1's in the corner cells are defined by saying that they are not the 0's, viz.:

$$f = \overline{B + D}$$

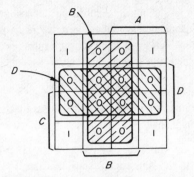

Fig. 3.17   Defining the 1's as NOT the 0's

It is left for the reader to show that the corner cells are also defined by the relationship $f = \bar{B} . \bar{D}$.

Example 3.1: Draw the Karnaugh map of the function

$$f = \bar{D}(\bar{A}[\bar{C} + \bar{B} . C] + A[\bar{C} + \bar{B} . C]) + B . \bar{C}. D$$

Simplify using the procedures outlined in this chapter.

Solution: The expression is first expanded as follows

$$f = \bar{A} . \bar{C} . \bar{D} + A . \bar{C} . \bar{D} + B . \bar{C} . D + \bar{A} . \bar{B} . C . \bar{D} + A . \bar{B} . C . \bar{D}$$

The expanded terms are mapped in Fig. 3.18, and adjacent 1's are grouped together in the simplest form possible, giving two solutions:

$$f = \bar{B} . \bar{D} + B . \bar{C}$$
$$= \overline{(B + D)} + B . \bar{C}$$

Alternatively the 0's in Fig. 3.18 are defined by $B . C + \bar{B} . D$, hence

$$f = \overline{B . C + \bar{B} . D}$$

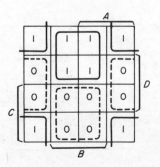

Fig. 3.18

Many problems have several minimal solutions, all being equally correct. In Example 3.1 three solutions were obtained. A network designer defines the *minimal* solution as one which gives the minimum number of circuit elements together with the minimum number of interconnections between the elements. Each solution obtained by the procedures outlined does not give the same number of interconnections.

The process of minimization of logical expressions by the map method is reduced to the following steps.

(a) Expand the original function to give logical product or sum terms.
(b) Map each term on the Karnaugh map by a series of 1's.
(c) Group the 1's into the largest blocks of cells possible (in binary groups).
(d) Write down the sum of the logical expressions obtained.

In step (c), the 0's could be grouped to give the complement of the final expression.

The advent of integrated electronic circuit technology has reduced the relative importance of minimization techniques, since complex circuits are obtainable in encapsulated form. Minimization techniques generally concentrate on producing circuits with the minimum number of integrated circuit packages, together with the minimum number of interconnections between them.

3.8   Maps for more than four variables

A map for two variables is obtained by putting two one-variable maps side by side, giving four cells in all. A three-variable map comprises two two-variable, or four one-variable maps side by side. A square array of two three-variable, or four two-variable maps gives a four-variable map.

The logical extension, for five variables, is to place two four-variable maps side by side, as shown in Fig. 3.19. The cell marked with a '1' in the E half is $A . B . \bar{C} . D . E$, and that in the $\bar{E}$ half is $A . B . \bar{C} . D . \bar{E}$. Giving the letters binary values, with A as the most significant digit, the cells have the values 11011 and 11010 respectively. Since the two numbers differ by only one binary

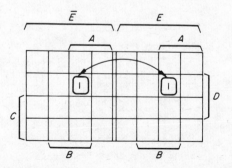

Fig. 3.19   One method of illustrating a five-variable map

digit, they are adjacent in accordance with the definition in section 3.5. If the $\bar{E}$ map is placed below the E map, then cells next to one another in the vertical plane are adjacent. In any one plane the normal rules of adjacency apply. An example is shown in Fig. 3.20 in which selected adjacent groups of cells are indicated.

The basic idea can be extended to give a six-variable map, which is two five-variable or four four-variable maps placed side by side in a square array. Adjacency between cells can be deduced by considering each equivalent four-variable map to be stacked one above the other. As in previous cases, adjacent cells must occur in multiples of two before they can be grouped together.

Usage of the map method becomes difficult when more than four variables are involved. One novel method for extending their use up to eight variables is described in Ref. 3. Alternative minimization techniques[4,5] include algebraic and tabular methods. These have the advantage that many more variables than four can be dealt with, but they lack the essential simplicity of map methods.

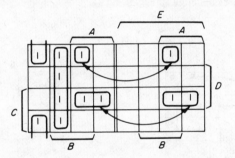

Fig. 3.20   Some adjacent cells on a five-variable map

3.9   Generating unit-distance codes on the Karnaugh map

Unit-distance codes (see sections 1.12 and 1.12.2) can readily be mapped on the Karnaugh diagram. A feature of unit-distance codes is that they change by only one binary digit for each consecutive group of code numbers. On the Karnaugh map this corresponds to the movement from one cell to an adjacent cell for a change from one code group to another. This is illustrated in Fig. 3.21 for a five-bit Gray code, where A is the most significant digit. The code commences at cell 00000 and moves through each adjacent cell in the $A = 0$ matrix, until it reaches cell $ABCDE = 01000$, when it transfers to the adjacent cell 11000 in the $A = 1$ matrix. The code follows the inverse pattern through this matrix, to the cell 10000, when it can either transfer to the start of the code (if it has a maximum of five bits), or it can enter the next matrix if there are six or more bits in the code. A four-bit code is generated if the pattern returns from cell 1000, on the $A = 0$ map, to cell 0000 on the same map.

It is evident that many forms of code can be generated by this technique. A

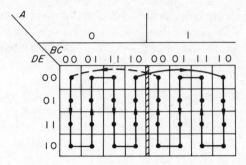

Fig. 3.21   Karnaugh map of a 5-bit Gray code

code which utilizes all the cells in a given matrix is known as a *complete cyclic code*, one example being shown in Fig. 3.22(a). A code which does not use all the cells in the matrix is an *incomplete cyclic code*, one example being the unit-distance decimal code in Fig. 3.22(b).

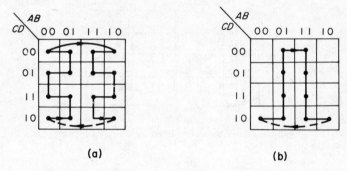

(a)   (b)

Fig. 3.22 (a) and (b)   Two forms of unit-distance code

Other forms of code such as the walking code described in section 1.10 can be traced out on the Karnaugh map.

3.10   Theorems and laws of logic

The truth of many logical statements is self-evident, while that of other statements may not be so clear. Providing that the statement is accurate, it is possible to test its truth using the ideas developed in this chapter.

    Using the binary notation, we say that a statement is true, i.e., the function exists, if it has the logical value '1'. If it is untrue, or does not exist, then it has the value '0'. The Karnaugh map can be used to prove logical theorems. Examples are shown in Figs. 3.23(a) and (b), which correspond to Theorems 1 and 2 below.

Theorem 1:	$A + 0 = A$
Theorem 2:	$A \cdot 0 = 0$
Theorem 3:	$A + 1 = 1$
Theorem 4:	$A \cdot 1 = A$
Theorem 5:	$A + A = A$
Theorem 6:	$A \cdot A = A$
Theorem 7:	$A + \bar{A} = 1$
Theorem 8:	$A \cdot \bar{A} = 0$
Theorem 9:	$\bar{\bar{A}} = A$

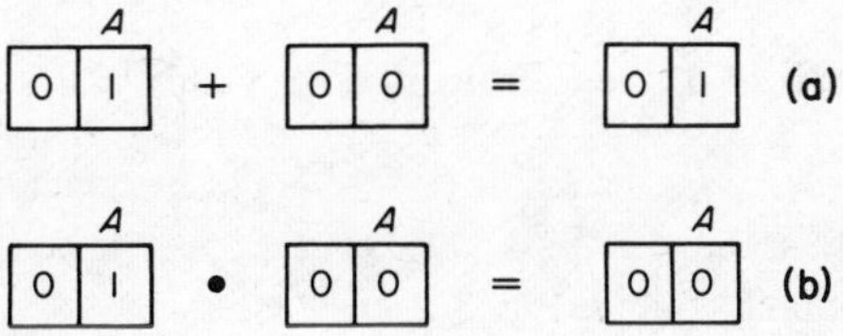

Fig. 3.23   Proof of theorems 1 and 2 are shown in (a) and (b), respectively

Certain laws are self-evident from the logical equations. Examples of these are the commutative and associative laws.

Commutative law

$$A + B = B + A$$
$$A \cdot B = B \cdot A$$

Associative law

$$A + B + C = (A + B) + C = A + (B + C)$$
$$A \cdot B \cdot C = (A \cdot B) \cdot C = A \cdot (B \cdot C)$$

Other theorems require further investigation, the distributive law being an example.

Distributive law

$$A + (B \cdot C \cdot D \ldots) = (A + B) \cdot (A + C) \cdot (A + D) \ldots$$
$$A(B + C + D + \ldots) = A \cdot B + A \cdot C + A \cdot D + \ldots$$

The second statement of the distributive law is self-evident, but the first statement requires further investigation. Its justification can be shown on a Karnaugh map, using the above techniques.

One of the most powerful tools in engineering applications is De Morgan's theorem, given below.

De Morgan's theorem

$$\overline{A + B + C + \ldots} = \bar{A} \cdot \bar{B} \cdot \bar{C} \ldots \tag{3.1}$$

$$\overline{A \cdot B \cdot C \ldots} = \bar{A} + \bar{B} + \bar{C} + \ldots \tag{3.2}$$

Equations (3.1) and (3.2) are proved, for two variables A and B, using Karnaugh maps in Figs. 3.24(a) and (b), respectively.

An illustrative example of the use of De Morgan's theorem is taken from section 3.7 in which the complement of the logical statement $A . B + A . C$ is required.

$$f = A . B + A . C = A(B + C) \tag{3.3}$$

or
$$\bar{f} = \overline{A(B + C)}$$

Applying eq. (3.2), with $(B + C)$ treated as a single term

$$\bar{f} = \bar{A} + \overline{(B + C)}$$

Application of eq. (3.1) to the right-hand term yields

$$\bar{f} = \bar{A} + \bar{B} . \bar{C} \tag{3.4}$$

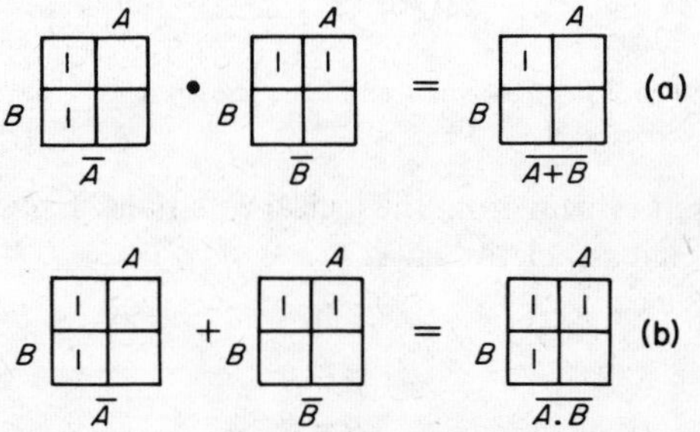

Fig. 3.24 (a) and (b)   Proof of De Morgan's theorem for two variables

De Morgan's theorem is expressed in general terms as follows. *The logical complement of a function is obtained by (1) logically inverting each term in the expression, and (2) by replacing the 'dots' with the 'plusses' and vice versa.*

For example, if

$$f = (A . \bar{B} . C) + (C . [A + \bar{D}]) + (E)$$

then
$$\bar{f} = (\bar{A} + B + \bar{C}) . (\bar{C} + [\bar{A} . D]) . (\bar{E})$$

It is advisable to collect each group of letters inside brackets to avoid mistakes. The brackets are not affected by the complementing process.

3.11   Canonical forms of expression

There are two normal or *canonical* forms of logical expression, namely *minterms* (terms which are ANDed) and *maxterms* (terms which are ORed). For example, the following Boolean expression is written entirely in minterm form.

$$f_1 = \bar{A} . \bar{B} . C + \bar{A} . B . C + A . B . C$$

The above expression is known as the *sum of products* (*S* of *P*) form. The following expression is written entirely in terms of maxterms.

$$f_2 = (A + B + \bar{C}) \,.\, (A + \bar{B} + \bar{C}) \,.\, (\bar{A} + \bar{B} + \bar{C})$$

It is also known as the *product of sums* (*P* of *S*) form of expression.

The above expressions can also be represented by means of decimal (or octal) values, using a 4, 2, 1 weighting for *A*, *B*, and *C*, respectively, as follows.

$$f_1 = \sum(1, 3, 7)$$

where the sigma symbol denotes the sum of products form. Also

$$f_2 = \prod(6, 4, 0)$$

where capital π denotes the product of sums form.

Problems

3.1. Plot Karnaugh maps of the functions $\overline{A \,.\, B \,.\, C}$, $\overline{A + B + C}$, $\overline{A + B + C}$, and $\bar{A} \,.\, \bar{B} \,.\, \bar{C}$. Hence show that the first and second functions are equivalent to one another, and that the third and fourth functions are equivalent to one another.

3.2. Write down the truth tables, and draw the Karnaugh maps of the following functions.

$$f_1 = A \,.\, B + \bar{A} \,.\, \bar{B} \qquad f_2 = A \,.\, \bar{B} + B \,.\, C$$
$$f_3 = A + A \,.\, B \,.\, C \qquad f_4 = A + B + \bar{A} \,.\, \bar{B}$$

3.3. Using the Karnaugh map, devise unit-distance (a) octal, (b) decimal, and (c) duodecimal codes not given in this chapter.

3.4. Minimize the following, using symbolic logic and the Karnaugh map techniques.

(a) $W \,.\, \bar{Y} \,.\, \bar{Z} + W \,.\, \bar{X} \,.\, \bar{Z} + W \,.\, Y \,.\, Z + W \,.\, X \,.\, Z$
(b) $X \,.\, \bar{Y} \,.\, \bar{Z} + W \,.\, X \,.\, \bar{Y} + \bar{X} \,.\, \bar{Y} \,.\, Z + X \,.\, Y \,.\, \bar{Z}$
(c) $W \,.\, X \,.\, Y \,.\, Z + \bar{W} \,.\, X \,.\, Z + X \,.\, \bar{Y} \,.\, Z + \bar{W} \,.\, \bar{X} \,.\, Y \,.\, \bar{Z} + W \,.\, \bar{X} \,.\, Z$

3.5 Minimize, using a Karnaugh map, the function
$$f = \bar{A} \,.\, \bar{B} \,.\, \bar{C} + \bar{B} \,.\, C \,.\, \bar{D} + \bar{A} \,.\, B \,.\, D + A \,.\, B \,.\, C \,.\, D + A \,.\, \bar{C} \,.\, D + A \,.\, \bar{B} \,.\, \bar{C} \,.\, \bar{D}.$$
Hence or otherwise show that it may be reduced to

$$f = \overline{B \,.\, \bar{D} + \bar{B} \,.\, C \,.\, D}$$

References

1. VEITCH, E. W., 'A chart method for simplifying truth functions', *Proc. Assoc. Comp. Mach.*, p. 127, May, 1962
2. KARNAUGH, M., 'The map method for synthesis of combinational logic circuits', *Commun. Electron.*, 9, 11, 539, 1953
3. DEAN, K. J., 'An extension of the use of Karnaugh maps in the minimization of logical functions', *J. Inst. Elect. Radio Engrs.*, 35, 5, 294–6, 1968
4. McCLUSKEY, E. J., 'Minimization of Boolean functions', *Bell Syst. Tech. J.*, 35, 6, 1417, 1956
5. PHISTER, M., *Logical design of digital computers*, John Wiley.

4. Representation of logical networks

It is desirable, although not essential, to understand the operation of the actual logic devices used in the solution of problems. Since many electronic, magnetic, fluid, and other logic devices are now used, a general treatment using block diagrams is given before a detailed study of the devices employed is undertaken.

4.1 Symbolic representation

Many conventions are at present employed to represent logic devices. To avoid confusion, the name of the device used is written inside the circle representing the logic gate. This is illustrated in Fig. 4.1 for NOT, AND and OR gates; a range of other logic symbols are given in the appendix.

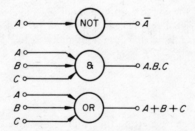

Fig. 4.1   Symbolic representation of NOT, AND, and OR gates

The term 'gate' is agricultural in origin since if a gate is open it is possible for livestock to pass freely through it. Logic elements are regarded as gates since there is flow of information when the gate is *open* and none when it is *closed* or *inhibited*.

48

4.2   Use of Karnaugh maps

It is useful to map functions in order to derive a block diagram of the logical network. An example is shown in Fig. 4.2 for the sum of two binary variables A

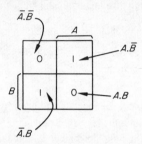

Fig. 4.2   Karnaugh map of the sum of two binary variables

and B. Table 4.1 lists the four possible combinations of A and B, and it is observed from both Fig. 4.2 and Table 4.1 that a sum occurs when we have $A . \bar{B}$ OR $\bar{A} . B$. That is,

$$S' = A . \bar{B} + \bar{A} . B$$

Table 4.1

Truth table for the sum and carry functions of two binary variables

A	B	Sum S'	Carry C'
0	0	0	0
1	0	1	0
0	1	1	0
1	1	0	1

The logical block diagram of this function is shown in Fig. 4.3, each term in the expression being generated separately. This circuit can also be used for the comparison of two binary digits. If $A > B$ (i.e., $A = 1$, $B = 0$) the output from

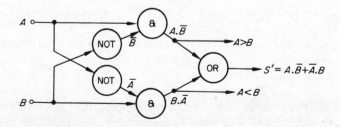

Fig. 4.3   Block diagram of a logic network which gives the sum of two binary variables. This network can also be used as a comparator for two variables

the upper AND gate is '1' while that from the lower AND gate is '0'. Similarly if $A < B$ ($A = 0, B = 1$) the lower AND gate gives a '1' while the upper AND gate output is '0'.

De Morgan's theorem can be used in conjunction with Fig. 4.2 to yield other logic networks. It is clear that the area marked by 1's is NOT the area marked by 0's, hence

$$S' = \overline{A} . \overline{B} + A . B = (\overline{\overline{A} . \overline{B}}) . \overline{A . B} = (\overline{\overline{A}} + \overline{\overline{B}}) . \overline{A . B}$$

$$= (A + B) . \overline{A . B}$$

This equation gives the network in Fig. 4.4. This circuit is more economic in terms of the number of logic gates required compared with Fig. 4.3, and has an additional advantage which is described later in this section. De Morgan's theorem can be applied again to give another network described by the function $(A + B) . (\overline{A} + \overline{B})$.

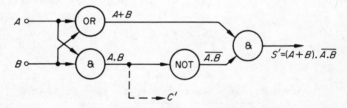

Fig. 4.4   Alternative block diagram of a network giving the sum of two binary variables

The logic networks developed so far give an output when the two inputs are not equivalent, i.e., if $A = 1, B = 0$, OR $A = 0, B = 1$. These networks are known as NOT-EQUIVALENT or EXCLUSIVE-OR gates, the circuit representation being shown in Fig. 4.5(a). The not-equivalent operation is also known as MODULO-2 ADDITION and the symbol $\oplus$ is used to express the function:

$$A \oplus B = \overline{A} . B + A . \overline{B}$$

The block diagram of the modulo-2 addition gate is given in Fig. 4.5(b). When a network with N inputs has to give a '1' output when an odd number of inputs have the value '1', a modulo-2 combination can be used. Examples of this are found in parity check networks and in feedback shift registers (chapter 7). A schematic diagram for this form of combination is given in Fig. 4.5(c).

Equivalence between two bits can be detected by using a network that gives zero output when $A \neq B$, and unity output when $A \equiv B$. This is satisfied by complementing or negating the output from a not-equivalent gate, as shown in Fig. 4.6. Alternatively a new series of logic networks can be derived by noting that when $A \equiv B$ the logical equation to be solved is $\overline{A} . \overline{B} + A . B$. The reader should justify this equation and derive suitable block diagrams.

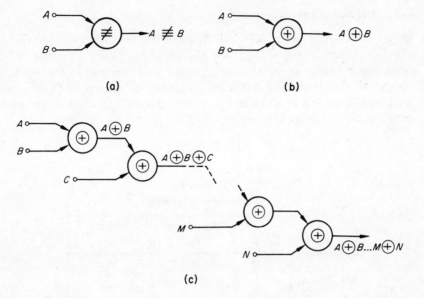

(a) (b)

(c)

Fig. 4.5 Symbolic representation of the NOT-EQUIVALENT or EXCLUSIVE-OR gate (a). The MODULO-2 addition version is shown in (b), and a network with N inputs is shown in (c)

Table 4.1 shows that a carry occurs when inputs A AND B occur simultaneously. A block diagram for the complete addition of two bits is shown in Fig. 4.7, which requires one additional AND gate to be added to Fig. 4.3. A feature of the network in Fig. 4.4 is that the logical product $A . B$ is developed at an early point, shown as C'. Figure 4.4 thus provides both sum and carry digits without further modification.

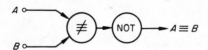

Fig. 4.6 Generation of the EQUIVALENT function

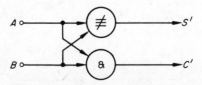

Fig. 4.7 Basic diagram of a network which gives the sum and carry functions of two binary variables

4.3 The full-adder

The circuit described in Fig. 4.7 is called a *half-adder* since it provides the sum and carry outputs corresponding to the addition of two bits. In a practical calculation it is necessary to account for the carry bit generated by the previous calculation. The process of full addition of two bits and a 'carry in' C_I from the previous addition is given in Table 4.2. An output sum S_O is seen to occur when an odd number of inputs are present, that is

$$S_O = A \oplus B \oplus C_I$$

Table 4.2

Truth table for a full-adder

A	B	C_I	Output sum S_O	Output carry C_O
0	0	0	0	0
0	0	1	1	0
0	1	0	1	0
0	1	1	0	1
1	0	0	1	0
1	0	1	0	1
1	1	0	0	1
1	1	1	1	1

The output carry equation is derived from the truth table as follows.

$$C_O = \bar{A} . B . C_I + A . \bar{B} . C_I + A . B . \bar{C_I} + A . B . C_I$$
$$= C_I(\bar{A} . B + A . \bar{B}) + A . B(\bar{C_I} + C_I)$$
$$= C_I . S' + A . B = C_I . S' + C'$$

The equations for S_O and C_O are combined in Fig. 4.8 to give one form of *full-adder* circuit.

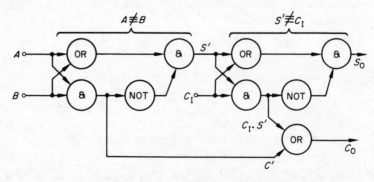

Fig. 4.8   One practical form of full-adder

4.4 NAND and NOR gates

Each element so far described performs only one function, e.g., the AND gate with given logic levels cannot be used to perform the NOT or OR functions. Two basic universal logic elements, the NOR and the NAND gates, are in common use today. Their functions are described in the following sections, and it will be shown that it is possible to construct any of the gates already described by suitable combinations of either type of universal logic element.

4.4.1 The NOR gate

The name NOR is derived from the logical statement

$$NOR = \overline{OR}.$$

That is, the NOR function is an inverted or negated OR function, illustrated in Fig. 4.9. The truth table for this function is developed in Table 4.3 for a two-input gate. It is seen that the output is zero if a '1' appears at either or both

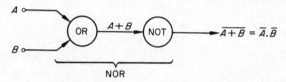

Fig. 4.9   Generation of the NOR function with discrete function elements

inputs, and that the output is '1' only when both inputs are zero. This can be extended to an n input NOR gate, since the output is '1' when all the inputs are 0's, otherwise it is zero.

It is uneconomic to use two gates, an OR and a NOT gate, to perform the NOR function, but in practice two gates are not necessary. Many electronic and fluid logic devices perform the basic NOR function with greater economy than they can carry out the OR and AND functions. This is discussed in more detail in later chapters.

Table 4.3

Truth table for a NOR gate with two inputs

Inputs		$A + B$	Output
A	B		$\overline{A + B}$
0	0	0	1
0	1	1	0
1	0	1	0
1	1	1	0

From De Morgan's theorem

$$\overline{A + B + C + \ldots} = \bar{A} \cdot \bar{B} \cdot \bar{C} \ldots$$

Fig. 4.10   The NOR gate with a single input acts as a NOT gate

Since it is frequently easier to think in terms of OR and AND gates, the output of the NOR gate may be considered as the AND function of the complements of the inputs to the NOR gate. It follows from this statement that if only input A exists (B, C, etc. $= 0$), then the output is $\bar{A}$, i.e., a NOR gate with a single input performs the NOT function. This is illustrated in Fig. 4.10. The logical OR function is generated by complementing the output of the NOR gate since

$$A + B + C + \ldots = \overline{\overline{A + B + C + \ldots}}$$

This is shown in Fig. 4.11.

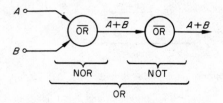

Fig. 4.11   Generation of the OR function with NOR elements

De Morgan's theorem allows the AND function to be realized with NOR gates since

$$\overline{A \cdot B} = \bar{A} + \bar{B}$$

or

$$A \cdot B = \overline{\bar{A} + \bar{B}}$$

The block diagram of the AND gate derived from NOR elements is shown in Fig. 4.12.

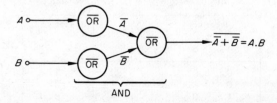

Fig. 4.12   The AND function is generated using three NOR gates

It may seem that any system built up of NOR gates will contain many more elements than are required using conventional discrete function gates, i.e., AND, OR, NOT gates. This is not necessarily so in practice since it is possible to eliminate many elements by inspection of the block diagram. One instance is shown in Fig. 4.13 where a single input A is applied to two cascaded NOR gates.

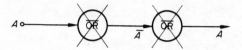

Fig. 4.13   Two cascaded NOR gates are both redundant

Both gates can be eliminated since the output is equal to the input. Another example is shown in Fig. 4.14(a), together with the simplified block diagram in Fig. 4.14(b). In Fig. 4.14(a) the output $\overline{A+B}$ from NOR 1 is complemented to $A + B$ by NOR 2. The final gate performs the NOR function on the two inputs, giving an output of $\overline{A + B + C}$. This output may be obtained from one NOR gate with inputs A, B, C as shown in Fig. 4.14(b).

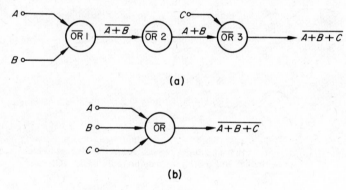

Fig. 4.14   Figure (a) can be minimized to (b)

4.4.2   Realization of NOR networks

If the statement of any problem can be expressed in the form of the logical product (AND) of a number of logical sums (OR's), then the OR and AND gates can be replaced directly with NOR gates.

This is illustrated in Fig. 4.15 for the function

$$f = (A + B) . (C + D)$$

The logic block diagram using discrete function elements is shown in Fig. 4.15(a). Using the combinations described in section 4.4.1, NOR equivalents of these elements are inserted giving Fig. 4.15(b). Using the example of Fig.

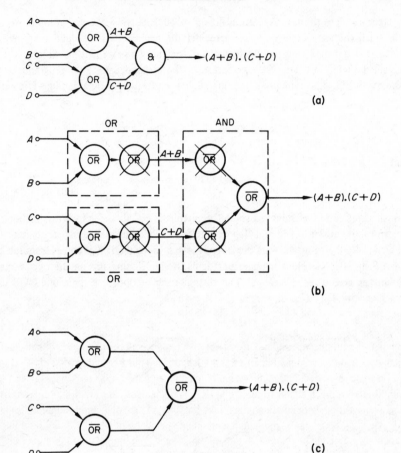

Fig. 4.15 (a), (b), and (c)   If the problem is presented in the form of the logical product of sums, the OR and AND gates may be replaced by NOR gates

4.13, four of the NOR gates are eliminated giving the final configuration in Fig. 4.15(c). It is evident that the OR and AND gates in Fig. 4.15(a) can be replaced by NOR gates.

This method does not always produce minimal networks, that is containing the minimum number of logic elements, but has the element of simplicity. An example is given below to illustrate the technique.

Example 4.1: Devise a NOR network to solve the following problem.

$$f = (A + \bar{B} + D) . (B + C + \bar{D}) . (\bar{A} + B + \bar{D})$$

Solution: The block diagram of the network is shown in Fig. 4.16, seven NOR gates being required, three for the purpose of generating $\bar{A}, \bar{B}, \bar{D}$. Using discrete function elements, three NOT, three OR, and one AND gates would be needed.

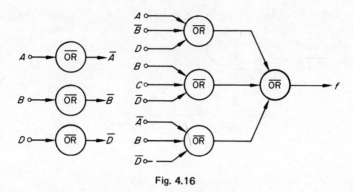

Fig. 4.16

4.4.3 A minimization method for NOR networks

Neither NOR nor NAND networks are always easy to minimize. Several ingenious methods have been suggested.[1,2] One method of network minimization is described here.

The output of a NOR gate with two inputs $\bar{A}, B$ is

$$f = A \cdot \bar{B} \quad \text{or} \quad \bar{f} = \bar{A} + B$$

Thus by inverting the output function and expressing it in OR logic form, the separate inputs are defined. In Fig. 4.17(a) $\bar{A}$ is generated by NOR 1. If that network is part of a larger network, it may be possible to use an alternative signal Z to replace the $\bar{A}$ input to NOR 2. This is illustrated in Fig. 4.17(b). In the following, a procedure to establish possible values of Z is developed.

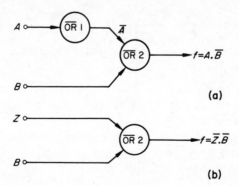

Fig. 4.17 Two circuits which are equivalent to one another; input $\bar{A}$ in (a) being replaced by Z in (b)

The truth table for Fig. 4.17(b) is given in Table 4.4. A more convenient form of this table is given in Table 4.5. It is observed that $\bar{Z}$ has the same value as f, except in the case when $\bar{B} = 0$, $f = 0$, when it is undefined, since it may be

Table 4.4

Truth table for Fig. 4.17(b)

$\bar{B}$	Z	$f = \bar{B} \cdot Z$
1	0	0
1	1	1
0	X	0

X = 'don't care', i.e., could be '0' or '1'.

Table 4.5

Evaluation of the $\bar{Z}$ function

$\bar{B}$	f	$\bar{Z}$
1	0	0
1	1	1
0	0	X

either '0' or '1'. Multiplication of the $\bar{B}$ and f Karnaugh maps, using Table 4.5, yields the $\bar{Z}$ map, illustrated in Fig. 4.18(a) for the network of Fig. 4.17(b). The Z map in Fig. 4.18(b) is obtained by converting the 1's to 0's and 0's to 1's in Fig. 4.18(a), the X's remaining unaltered. The resulting logical expressions for Z, from Fig. 4.18(b), are $\bar{A}, \bar{A} \cdot B$, or $\bar{A} + B$. If any of these are generated at any point within the network, NOR 1 becomes redundant.

The general method of applying the network minimization technique is now described.

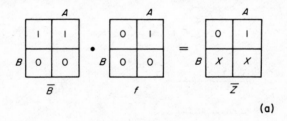

(a)

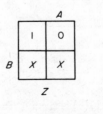

(b)

Fig. 4.18   The method of evaluating the $\bar{Z}$ function in (a) utilizes table 4.5. The Z function is given in (b)

1. Obtain the complement of the desired function to give the inputs to the final NOR gate. Common factors should be grouped together to give all possible forms of input.
2. Generate each term in resulting expressions separately.
3. Inspect each circuit for redundant NOR gates. The network is completed by feeding a final NOR gate from the minimal circuits obtained. If there are several networks with equal numbers of gates, the minimal network is the one with the least number of interconnections.

Example 4.2: Design a NOR network to give an output when $A \equiv B$.

Solution:

Step 1

$$f = A \cdot B + \bar{A} \cdot \bar{B}$$

The function is mapped in Fig. 4.19(a) and complemented in (b), yielding

$$\bar{f} = \bar{A} \cdot B + A \cdot \bar{B}$$

(a)   (b)

Fig. 4.19 Karnaugh map (a) of f, and (b) of $\bar{f}$ in example 4.2

Step 2 The functions $\bar{A} \cdot B$ and $A \cdot \bar{B}$ are generated separately in Figs. 4.20(a) and (b) respectively.

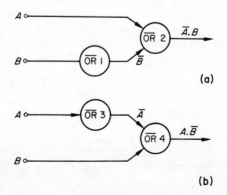

Fig. 4.20 Generation of (a) $\bar{A} \cdot B$ and (b) $A \cdot \bar{B}$ for example 4.2

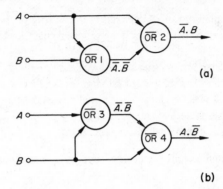

Fig. 4.21 (a) and (b)   A method of eliminating one NOR gate

Step 3 NOR 1 and NOR 3 gates are now investigated for redundancy. In the event that either or both cannot be eliminated, the outputs of NOR 2 and NOR 4 will be fed to another NOR gate, the output from which is the desired function. Using the technique described above, Figs. 4.20(a) and (b) are re-drawn in Figs. 4.21(a) and (b) respectively. The output $\bar{A} . \bar{B}$ can be used as a common input to gates NOR 2 and NOR 4, reducing the number of gates required by one. The minimal form of the final network is shown in Fig. 4.22.

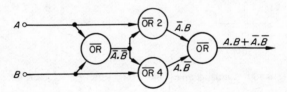

Fig. 4.22   The minimal NOR network for the solution of example 4.2

4.4.4 A minimization method for NOR gates with three inputs

In most combinational logic it is usual to have gates with more than two inputs. The method of extending the above treatment is given here. A network, together with its Karnaugh map is shown in Fig. 4.23. If Z is the input to NOR 2 which allows NOR 1 to be eliminated, the output from NOR 2 is $\bar{Z} . \bar{B} . \bar{C}$. The $\bar{B} . \bar{C}$

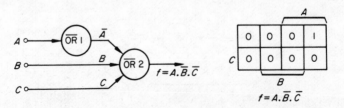

Fig. 4.23   A NOR gate with three inputs, one signal being inverted

component of the output is essential since there are no NOR gates in the B and C input lines. The $\bar{Z}$ map is thus generated by multiplying the $\bar{B}$. $\bar{C}$ map with the f map in accordance with Table 4.5, where the $\bar{B}$ function in Table 4.5 is replaced by the $\bar{B}$. $\bar{C}$ in this example. The result of this logical product is shown in Fig. 4.24(a) and its inverse, the Z map, in Fig. 4.24(b).

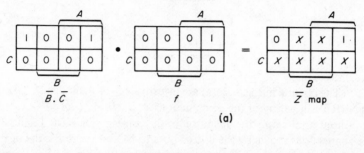

(a)

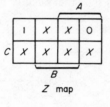

Z map

(b)

Fig. 4.24   Generation of (a) the $\bar{Z}$ and (b) the Z function for Fig. 4.23

Any combination which describes the cell $\bar{A}$. $\bar{B}$. $\bar{C}$ but not the cell A. $\bar{B}$. $\bar{C}$ will act as a satisfactory input. These include $\bar{A}, \bar{A}$. $\bar{B}, \bar{A}$. $\bar{C}, \bar{A}$. $\bar{B}$. $\bar{C}, \overline{A}. \overline{B}. \overline{C}$ $\bar{A}(\bar{B} + C), \bar{A}(B. \bar{C})$, etc.

Example 4.3: Derive a minimal NOR network for the function

$$f = A . B + B . C + A . C + \bar{A}. \bar{B}. \bar{C}$$

Solution:

Step 1   From Fig. 4.25

$$\bar{f} = A . \bar{B}. \bar{C} + \bar{A}. B . \bar{C} + \bar{A}. \bar{B}. C \qquad (4.1)$$
$$= \bar{A}(B. \bar{C} + \bar{B}. C) + A . \bar{B}. \bar{C} \qquad (4.2)$$
$$= \bar{B}(A . \bar{C} + \bar{A}. C) + \bar{A}. B . \bar{C} \qquad (4.3)$$
$$= \bar{C}(A . \bar{B} + \bar{A}. B) + \bar{A}. \bar{B}. C \qquad (4.4)$$

Step 2   Each term in each of the above equations has to be generated separately in order to determine the minimal network. Equations (4.1) and (4.2) are dealt with in detail here, Figs. 4.26(a) and (b) corresponding to eqs. (4.1) and (4.2) respectively. It is left to the reader to show if eqs. (4.3) and (4.4) give networks more minimal than that in the final Fig. 4.27.

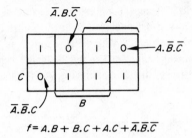

$$f = A.B + B.C + A.C + \overline{A}.\overline{B}.\overline{C}$$

Fig. 4.25   Karnaugh map for example 4.3

Step 3   The circuits in Fig. 4.26(a) are investigated for redundancy. The output from NOR 1 can be any of the combinations $\overline{A}, \overline{A}.\overline{B}, \overline{A}.\overline{C}, \overline{A}.\overline{B}.\overline{C}, \overline{A}(\overline{B}+C)$, etc., without invalidating the operation of the combination NOR 1 and NOR 2. If the connections shown by the dotted lines to NOR 1 are made, the output of NOR 1 is $\overline{A}.\overline{B}.\overline{C}$. Investigation of the other circuits in Fig. 4.26(a) shows that

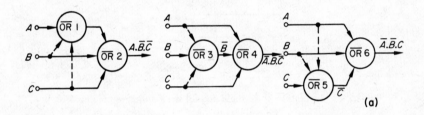

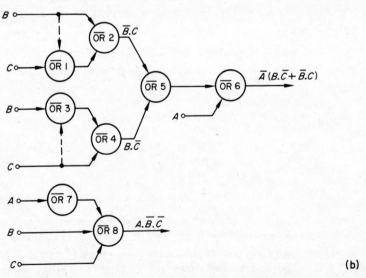

Fig. 4.26   (a) Networks for equation (4.1), and (b) networks for equation (4.2)

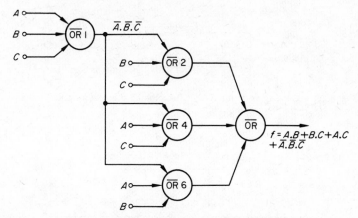

Fig. 4.27 Minimal network for equation (4.1)

the signal $\bar{A}\,.\,\bar{B}\,.\,\bar{C}$ can replace the $\bar{B}$ output from NOR 3 and the $\bar{C}$ output from NOR 5 (the reader should verify this). Thus NOR 3 and NOR 5 can be eliminated.

In Fig. 4.26(b) the combination $\bar{A}(B\,.\,\bar{C}+\bar{B}\,.\,C)$ is realized. Investigation of Fig. 4.26(b) shows that NOR 3 can be eliminated if the dotted connection to NOR 1 is made. The output $\bar{B}\,.\,\bar{C}$ can be used as an input to NOR 2 in place of $\bar{C}$, and to NOR 4 in place of $\bar{B}$.

The networks in Fig. 4.26(a) are the minimal combinations, since only four NOR gates are required to generate all the terms in eq. (4.1). In Fig. 4.26(b) seven gates are required to generate all in terms in eq. (4.2). The final combination of five NOR gates is shown in Fig. 4.27.

4.4.5   A NOR gate with N inputs

In general, for a NOR gate with N inputs $A, B, C, \ldots, L, M, N$, with the A input line complemented by a NOR gate, the Karnaugh map for an input Z, which replaces the $\bar{A}$ line, has a '1' in the $\bar{A}\,.\,\bar{B}\,.\,\bar{C}\,.\,\ldots\,.\,\bar{L}\,.\,\bar{M}\,.\,\bar{N}$ cell, and a '0' in the $A\,.\,\bar{B}\,.\,\bar{C}\,.\,\ldots\,.\,\bar{L}\,.\,\bar{M}\,.\,\bar{N}$ cell. All the other cells are not defined. The required input Z, to make the NOR gate in the A channel redundant, must define the cell $\bar{A}\,.\,\bar{B}\,.\,\bar{C}\,.\,\ldots\,.\,\bar{L}\,.\,\bar{M}\,.\,\bar{N}$ but not the cell $A\,.\,\bar{B}\,.\,\bar{C}\,.\,\ldots\,.\,\bar{L}\,.\,\bar{M}\,.\,\bar{N}$.

4.4.6   The NAND gate

The name NAND is derived from the statement

$$\text{NAND} = \overline{\text{AND}}$$

that is, the NAND function is an inverted AND function. This is illustrated in Fig. 4.28 together with its truth table, Table 4.6. A NAND gate with N inputs gives a '0' output when all the inputs are '1', otherwise the output is '1'.

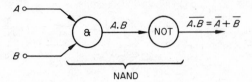

Fig. 4.28   Generation of the NAND function by discrete function elements

Table 4.6

Truth table of the NAND function

Inputs			Output
A	B	$A \cdot B$	$\overline{A \cdot B}$
0	0	0	1
0	1	0	1
1	0	0	1
1	1	1	0

From De Morgan's theorem

$$A \cdot B \cdot C \ldots = \bar{A} + \bar{B} + \bar{C} + \ldots$$

The output of the NAND gate may be considered as the OR function of the complements of the inputs to the NAND gate. If only one input line is used, the output is $\bar{A}$. That is, a NAND gate with a single input performs the NOT function, as shown in Fig. 4.29.

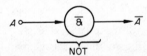

Fig. 4.29   The NAND gate with a single input acts as a NOT gate

De Morgan's theorem allows the OR gate to be realized:

$$\overline{A + B} = \bar{A} \cdot \bar{B}$$

or

$$A + B = \overline{\bar{A} \cdot \bar{B}}$$

The logical block diagram of this function is shown in Fig. 4.30.

As with the NOR gate, two cascaded NAND gates with a single input can be eliminated, and a combination of NAND gates, similar to the NOR network in Fig. 4.14(a) can also be reduced to one NAND gate.

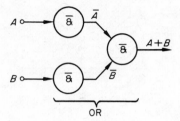

Fig. 4.30   The OR function is generated using three NAND gates

4.4.7   Realization of NAND networks

If the problem can be expressed in the form of the logical sum (OR) of a number of logical products (AND's), then the OR and AND gates can be replaced directly with NAND elements.

Consider the function

$$f = A \cdot B + C \cdot D$$

The logic network to solve this equation using discrete function elements is shown in Fig. 4.31(a). These are replaced in Fig. 4.31(b) by the NAND equivalents. Eliminating four of the gates leaves the final logic diagram in Fig. 4.31(c).

Using this technique it is first necessary to manipulate the equation of the problem into the form of the logical sum of products. As an example consider the statement

$$f = (A + B) \cdot (C + D)$$

The function is first complemented to give

$$\bar{f} = \bar{A} \cdot \bar{B} + \bar{C} \cdot \bar{D}$$

or

$$f = \overline{\bar{A} \cdot \bar{B} + \bar{C} \cdot \bar{D}}$$

giving the NAND network in Fig. 4.32.

4.4.8   Minimization of NAND networks

A similar technique to that outlined for NOR networks in section 4.4.3 can be developed for NAND networks. Where a NAND gate has N inputs $\bar{A}, B, C, \ldots, L, M, N$, the NAND gate in the A input line can be eliminated if a signal Z is available which defines the cell $\bar{A} \cdot B \cdot C \cdot \ldots \cdot L \cdot M \cdot N$, but not the cell $A \cdot B \cdot C \cdot \ldots \cdot L \cdot M \cdot N$ in the Karnaugh map. All other cells in the Z map contain X's.

The general minimization procedure is:

(a) Obtain the complement of the desired function.
(b) Generate each term in the resulting expression separately.

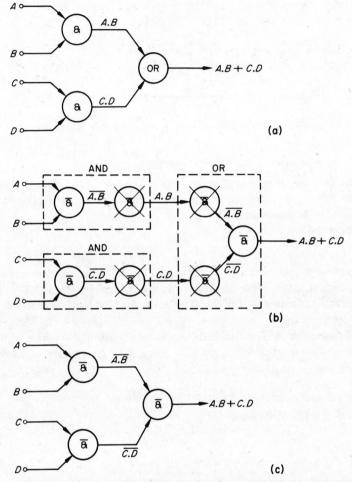

Fig. 4.31 (a), (b), and (c)   If the problem is presented in the form of the logical sum of products, the AND and OR gates may be replaced by NAND gates

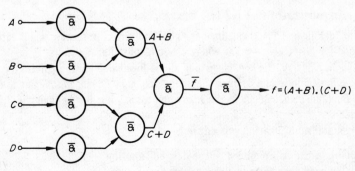

Fig. 4.32

(c) Inspect each circuit for redundant NAND gates. The network is completed by feeding a final NAND gate from the minimal input circuits.

FEATURES OF SIMPLIFICATION TECHNIQUES

The minimization techniques outlined in sections 4.4.3 and 4.4.8 are useful when dealing with systems of limited complexity, as also is the inhibiting loop technique.[1] However, these methods have a number of disadvantages, including the following:

1. A variable may be applied at several points in the network (see, for example, variable A in Fig. 4.27), and the differing path lengths through which the signal has to travel may give rise to hazards in the system (see also section 4.10).
2. In general, the saving in the number of gates is not very great. With the advents of ICs, minimization techniques concentrate on the saving of complete IC packages.
3. Servicing the minimized network may be made more difficult, since the logic function it generates may be masked by the apparent complexity of the circuit.

In many cases, it is advisable to use the simplest possible design techniques since these overcome some of the disadvantages listed above. Examples of simple design techniques are given in sections 4.5 to 4.8 below.

4.5 Realization of NAND networks directly from the Karnaugh map

Modern logic systems utilize a common type of gate, which may either be NAND or NOR gates. In this section of the book, a design procedure is outlined which allows NAND networks to be designed directly from Karnaugh maps. The networks produced in this way are not necessarily minimal networks. The steps involved in the design procedure are listed below:

1. Draw the Karnaugh map of the function and group the 1's in the manner outlined earlier, each loop being expressed in a logical product form, i.e., in the form $A . B$, etc.
2. Draw a two-tier NAND network having as many NAND gates in the first tier as there are loops of 1's on the Karnaugh map. The final tier contains only one NAND gate, the output from each gate in the first tier being used as an input to the second tier gate.
3. The variables defining one of the loops of 1's on the Karnaugh map are used as input signals to one of the first tier gates. This process is repeated for each loop on the Karnaugh map.
4. Additional NAND gates may be required to invert some of the input signals.

Consider the design of a NAND network which satisfies the logical equation

$$f = A . B . C + \bar{A} . B . \bar{C} + A . B . \bar{C} + \bar{A} . \bar{B} . C$$

The steps outlined above are illustrated in Fig. 4.33. Firstly the Karnaugh map
for the function is drawn, see Fig. 4.33(a), in which the groups of cells $A . B$,
$B . \bar{C}$ and $\bar{A} . \bar{B} . C$ are defined. A two-tier NAND network comprising gates
G1–G4, inclusive, is drawn in Fig. 4.33(b). Gate G1 is taken in association with
loop $A . B$ on the Karnaugh map, gate G2 in association with loop $B . \bar{C}$, and G3
in association with loop $\bar{A} . \bar{B} . C$. Accordingly, the two input lines of G1 are
energized by signals A and B, the input lines of G2 are energized by signals B and
$\bar{C}$, and signals $\bar{A}, \bar{B}$ and C are applied to G3. Three additional invertors – G5, G6
and G7 – are required to generate the functions $\bar{A}, \bar{B}$ and $\bar{C}$, respectively.
Readers may find it instructive to verify that the circuit in Fig. 4.33(b) generates
the original logic function.

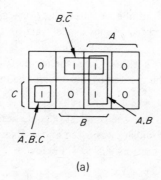

(a)

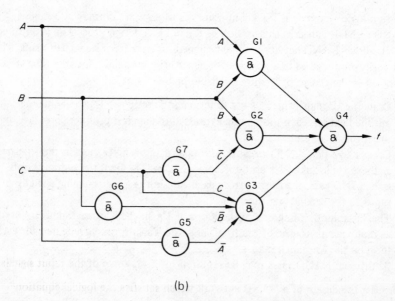

(b)

Fig. 4.33 (a) and (b)   One method of designing a NAND network from the Karnaugh map

4.6 Realization of NOR networks directly from the Karnaugh map

The procedure outlined below leads to a two-tier NOR network directly from the Karnaugh map. The steps to be followed are:

1. Draw the Karnaugh map, and group the 0's in the largest possible combinations.
2. Draw a two-tier NOR network having as many NOR gates in the first tier as there are loops of 0's on the Karnaugh map. The final tier contains only one NOR gate, the output from each gate in the first tier being used as an input to the second tier gate.
3. The *logical compliment* of the variables defining one of the loops on the Karnaugh map are used as input signals to one of the first tier gates. This process is repeated for each loop on the Karnaugh map.
4. Additional NOR gates may be required to invert some of the input variables.

To illustrate the above procedure, a design for a NOR network satisfying the Boolean function below is developed.

$$f = A \cdot B \cdot C + \bar{A} \cdot B \cdot \bar{C} + A \cdot B \cdot \bar{C} + \bar{A} \cdot \bar{B} \cdot C$$

The Karnaugh map for the function is drawn up in Fig. 4.34(a) (readers will note that it is identical to the map in Fig. 4.33(a)), and the 0's are looped in the most economical manner. Since there are three groups of 0's on the Karnaugh map, the two-tier NOR network in Fig. 4.34(b) contains three gates (G1–G3) in the first tier. Gate G1 in this tier is concerned with loop $A \cdot \bar{B}$ on the Karnaugh map and, from rule (3) above, this gate has two input signals which are $\bar{A}$ and $\bar{\bar{B}} = B$. This procedure is repeated for gates G2 and G3 in the figure. Finally, invertors G5, G6 and G7 are introduced to generate the functions $\bar{A}$, $\bar{B}$ and $\bar{C}$, respectively.

4.7 The WIRED-OR connection or distributed logic

A useful feature of certain types of electronic gates is that their outputs may be wired directly together. Depending on the type of gate, this feature either (a) causes a completely new logic function to be generated, or (b) leaves the function generated by the gates unchanged, but allows a greater number of input signals to be accommodated (this is known as increasing the *fan-in* of the gates). In general terms, it is possible to connect the outputs together in this way only if the output resistance of the gates is low when the output signal is at one of the logic levels (say logic '0'), and is high when the complementary output signal is present (logic '1'). Readers are cautioned that the WIRED-OR connection cannot be used with certain types of gate, such as conventional designs of TTL (see section 9.5).

Expressed in simple terms, *the WIRED-OR connection of a number of gates generates the logic AND function of the individual functions* of each of the

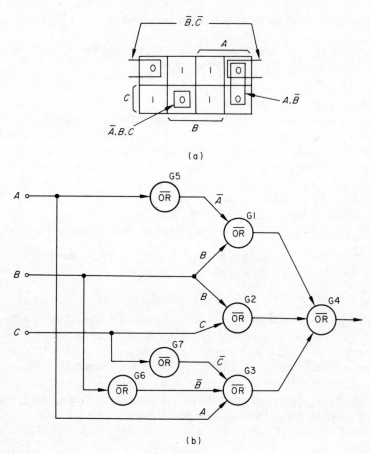

(a)

(b)

Fig. 4.34 (a) and (b)   One method of designing a NOR network from the Karnaugh map

gates. There is, of course, no objection to wiring gates together in this way which generate different types of logic function. For example, if the lower gate in Fig. 4.35 is replaced by a NAND gate, the function generated by the combination is $\overline{(A + B)} . \overline{A . B}$.

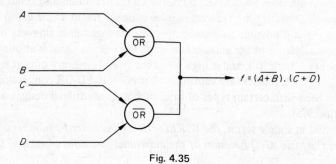

Fig. 4.35

4.7.1 Distributed logic NOR networks

When two NOR gates are connected in the manner shown in Fig. 4.35, the expression for the logical output from the combination is

$$f = \overline{(\overline{A+B})} \cdot \overline{(\overline{C+D})} = \overline{A} \cdot \overline{B} \cdot \overline{C} \cdot \overline{D} = \overline{A+B+C+D}$$

That is, by connecting NOR gates together in this way the overall function is unchanged, but the circuit can accommodate a greater number of input signals than is possible in the case of a single gate, i.e., the fan-in of the system is greater than that of a single gate.

Note: The above expression is true in networks employing gates in which the output impedance is low for a '0' output, and is high for a '1' output. If the reverse is the case, the resulting expression for the output is

$$f = \overline{(\overline{A+B})} + \overline{(\overline{C+D})}$$

Similarly for 4.7.2 below.

4.7.2 Distributed logic NAND networks

If the gates in Fig. 4.35 are replaced by two 2-input NAND gates, the logical expression for the output is

$$f = \overline{(\overline{A \cdot B})} \cdot \overline{(\overline{C \cdot D})} = (\overline{A} + \overline{B}) \cdot (\overline{C} + \overline{D})$$

This is a new logic function which differs from the basic NAND function.

4.8 Realization of WIRED-OR NAND networks directly from the Karnaugh map

The WIRED-OR connection of NAND gates generates a new type of function which leads, in some cases, to circuit simplification. An algorithm for designing WIRED-OR networks is given below:

1. Draw the Karnaugh map for the function, and group the 0's using the largest possible combinations.
2. Draw a WIRED-OR NAND network having as many NAND gates as there are loops on the Karnaugh map.
3. The input signals applied to a gate dealing with a specific loop on the Karnaugh map are given by the variables which define the loop on the map.
4. Additional NAND gates may be required to invert some input signals.

To illustrate the above procedure, a design is carried out for the Karnaugh map in Fig. 4.34(a). On that map, the loops joining the 0's were defined by the expressions $A \cdot \overline{B}, \overline{B} \cdot \overline{C}$ and $\overline{A} \cdot B \cdot C$. Rule (2) above indicates that, in this case, three NAND gates are required in the network, whose inputs (rule (3)) are A and $\overline{B}, \overline{B}$ and $\overline{C}$, and $\overline{A}$ and B and C, respectively. The resulting logic block diagram for this network is shown in Fig. 4.36. Additionally, three more NAND gates are required in order to generate the complements of variables A, B and C.

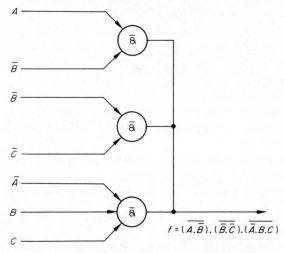

Fig. 4.36   Designing a WIRED-OR network

4.9   Logic signal levels

For most practical purposes it is convenient to think in terms of a true zero value, e.g., a zero value of voltage, current, pneumatic pressure, etc., as logical '0', and a finite positive value for logical '1'. This is known as *positive logic notation*, since the more positive of the two levels is '1'. Many early semiconductor logic devices gave an output voltage which was either zero or a negative value. For convenience zero voltage was called logical '0', and the finite negative voltage was logical '1'. This is known as *negative logic notation* since the more negative voltage of the two logic levels is taken as the logical '1'.

Many devices used today, both pneumatic and electronic, use 'floating' levels, e.g., one of the levels in a pneumatic system may be 20 kN/m^2 which the other may be 70 kN/m^2. If the higher of the two levels is taken as '1', then the device operates with positive logic. If the lower of the two levels is taken as '1', the device operates with negative logic. Illustrative examples of both positive and negative logic levels are shown in Figs. 4.37(a) and (b), respectively.

It is convenient in some instances to use mixed logic. For example the input logic could be positive logic while the output logic may be negative logic. That is a finite signal at the input represents a '1', while a signal at the output represents a '0'.

Conversion from positive logic to negative logic and vice versa can readily be carried out since

Positive logic = NOT negative logic
Negative logic = NOT positive logic.

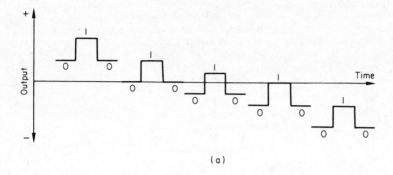

(a)

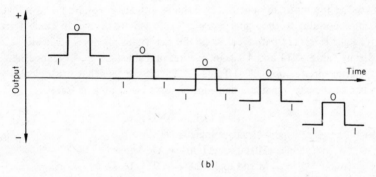

(b)

Fig. 4.37   Examples of (a) positive logic signal levels, and (b) negative logic levels

That is, either one is the logical complement of the other. If the input to a NOT gate is '1' in positive logic, the output may be regarded as '0' in positive logic or '1' in negative logic.

For the purposes of the following general discussion the actual output signal will be given, H representing high voltage and L low voltage. In positive logic $H = 1$, $L = 0$, and in negative logic $H = 0$, $L = 1$. The logical properties of AND, NAND, OR, and NOR gates are given in Table 4.7, the logical properties being independent of the logic level used.

Table 4.7

Properties of logic gates

Inputs		Outputs			
A	B	AND	NAND	OR	NOR
0	0	0	1	0	1
1	0	0	1	1	0
0	1	0	1	1	0
1	1	1	0	1	0

Table 4.8

Properties of the gate in section 4.5

Inputs		Output
A	B	
L	L	L
H	L	L
L	H	L
H	H	H

Consider a gate which is found to satisfy Table 4.8. It will be shown that the logical properties, and therefore the name of the gate, are dependent on the logic levels used at the input and output. In Table 4.9 the appropriate positive logic levels corresponding to the signal levels in Table 4.8 are shown. In Table 4.10 the negative logic levels are indicated, while the two types of mixed logic are recorded in Tables 4.11 and 4.12. Inspection of Tables 4.9 to 4.12, together with the appropriate columns of Table 4.7, yields the fact that the logical properties of the gate are dependent on the logic levels used as follows:

Positive logic — AND gate
Negative logic — OR gate
Mixed logic (1) — NAND gate
Mixed logic (2) — NOR gate

That is, the gate may be used to represent any logic function if the input and output logic levels are chosen correctly.

This state of affairs can be represented by the device matrix in Fig. 4.38, for the properties listed in Table 4.8. It is only possible to correctly identify a logic gate if both its logic function, e.g., AND, OR, etc., and the input and output logic levels are defined. To avoid confusion in later sections of this book, positive logic is assumed, unless otherwise stated.

The device matrix may be used to identify the logical properties of any gate with any input and output logic levels as follows. The four cells are assigned the

Table 4.9

Positive logic equivalent
of Table 4.8

Inputs		Output
A	B	
0	0	0
1	0	0
0	1	0
1	1	1

Table 4.10

Negative logic equivalent
of Table 4.8

Inputs		Output
A	B	
1	1	1
0	1	1
1	0	1
0	0	0

Table 4.11

Mixed logic (1) equivalent of
Table 4.8. Positive input
logic, negative output logic

Inputs		Output
A	B	
0	0	1
1	0	1
0	1	1
1	1	0

Table 4.12

Mixed logic (2) equivalent of
Table 4.8. Negative input
logic, positve output logic

Inputs		Output
A	B	
1	1	0
0	1	0
1	0	0
0	0	1

names given in Fig. 4.38, and the position of the device is located by the known values of input and output logic levels. All other logical properties are then defined accordingly by the matrix.

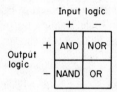

Fig. 4.38   The device matrix for the element described by Table 4.9

Example 4.4: A gate is found to have the logical properties of an OR gate when positive logic is used at the input, and negative logic at the output. If it is used with (a) positive logic, (b) negative logic, and (c) negative input and positive output logic, what logical functions are performed?

Solution: The names of the logic gates are located in the matrix in Fig. 4.39(a), then the appropriate input logic level is assigned to the column, and output logic level to the row. The remaining signs for input and output logic are then given to the other row and column in Fig. 4.39(b). The results are tabulated from the device matrix in Table 4.13.

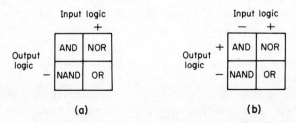

(a)   (b)

Fig. 4.39 (a) and (b)   The device matrix for example 4.4

Table 4.13

Logical properties of the gate in Example 4.4

	Input logic	Output logic	Logical function
	+	−	OR
(a)	+	+	NOR
(b)	−	−	NAND
(c)	−	+	AND

Example 4.5: Derive block diagrams showing how AND and OR functions can be generated using NOR gates.

Solution: Assuming that the NOR gate operates with positive logic, the device matrix is as shown in Fig. 4.39(b). The AND function is generated if the input logic is inverted, i.e., all input lines are fed through NOT gates (single input NOR gates). The block diagram is therefore as shown in Fig. 4.12. The OR function is obtained by inverting the output logic, i.e., the output from the NOR gate is fed through a NOT gate (single input NOR gate). This gives the configuration in Fig. 4.11.

4.10   Static hazards

In combinational logic the change of a single variable from '0' to '1', or '1' to '0', may cause a transient change in the output of a network when no change should exist. This is known as a *static hazard*.[3]

Two simple forms of static hazard are illustrated in Fig. 4.40. It is assumed in most logic devices that when $A = 1$, then $\bar{A} = 0$, and when $A = 0$, then $\bar{A} = 1$. This is true under steady-state conditions, but during transition periods the condition $A = \bar{A} = 1$, or $A = \bar{A} = 0$ may exist, as shown in the figure. This is due to the time taken for the signal to propagate through the NOT gate. Only if the propagation time is zero does $A + \bar{A} = 1$ and $A \cdot \bar{A} = 0$ at all times, eliminating the possibility of a static hazard in Fig. 4.40.

Due to the finite propagation time of practical gates, output $A + \bar{A}$ falls to zero for a short time and $A \cdot \bar{A}$ assumes the value '1' for an instant. In many networks this is of no consequence, but if the output of the combinational logic feeds a circuit which counts pulses, the spurious pulses produced by the static hazards may be counted in addition to the wanted pulses. This results in an inaccurate count.

It is possible to eliminate static hazards by including redundant gates in the network, but before these additional gates are included it should first be determined if the input conditions corresponding to the static hazard can occur.

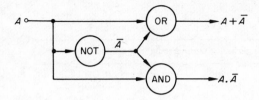

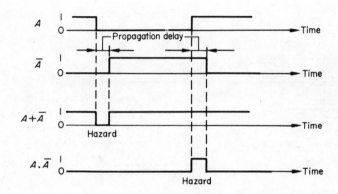

Fig. 4.40   Examples of static hazards

If this is not the case, then no static hazard occurs. Circuits can be designed to include known static hazards and still operate in a satisfactory manner.

A further example of a static hazard is considered here to illustrate the procedure for eliminating them. Consider the function

$$f = A \cdot C + B \cdot \bar{C}$$

Two networks are derived, using the Karnaugh map technique, in Figs. 4.41(a) and (b). Investigation of these networks shows that static hazards occur in both cases, but under different circumstances, as shown in Fig. 4.41(c). A hazard is said to exist when the input signal combinations change in such a way that they cause a change between adjacent cells in the Karnaugh map which are not grouped together. This is illustrated in Figs. 4.41(a) and (b) by the cells connected by arrows. By grouping these cells, as shown in Figs. 4.42(a) and (b), the hazards are eliminated. The resulting hazard-free networks are also shown in these figures.

An illustration of a four-variable problem in which hazards may exist is shown in Fig. 4.43.

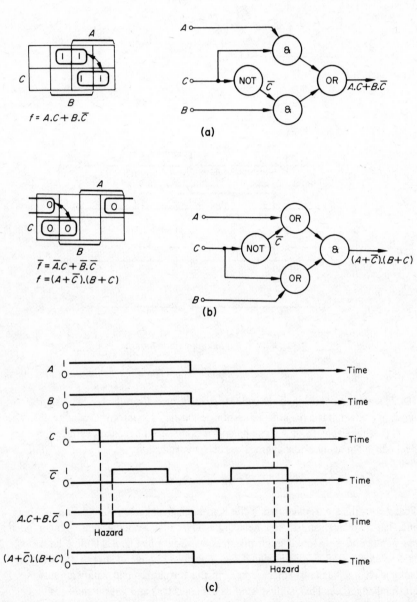

Fig. 4.41   Static hazards exist if the input signals change in such a way that they cause a change between adjacent cells on the Karnaugh map which are not grouped together. This is shown in (a) and (b) by the groups of cells linked by arrows. The hazards are illustrated in (c)

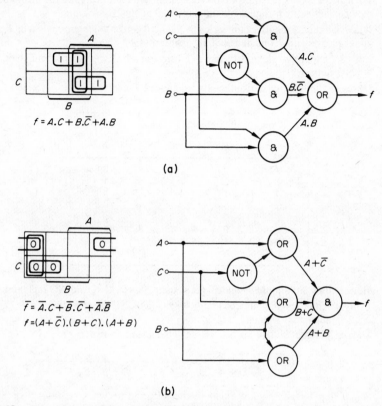

(a)

(b)

Fig. 4.42   The circuits in (a) and (b) are the hazard-free versions of those in Figs. 4.41(a) and (b), respectively

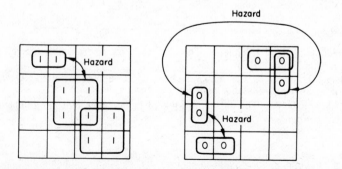

Fig. 4.43   Two four-variable problems in which static hazards may exist

Problems

4.1. With the aid of a truth table, deduce the logic equations for the sum and carry functions produced by the addition of two binary digits.

4.2. Draw a logic block diagram, using AND, OR and NOT elements, which generate the functions deduced in problem 4.1.

4.3. Convert the network in problem 4.2 into (a) a NOR network, and (b) a NAND network.

4.4. Develop the logic equation for the following truth table.

Inputs			Output
A	B	C	f
0	0	0	0
0	0	1	1
0	1	0	1
0	1	1	0
1	0	0	0
1	0	1	1
1	1	0	1
1	1	1	0

Hence draw up a logic block diagram using AND, OR, and NOT elements.

4.5. Convert the network in problem 4.4 into (a) a NOR network, and (b) a NAND network.

4.6. Devise minimal NOR and NAND networks for the functions

$$f_1 = \bar{A} \cdot \bar{B} \cdot \bar{C} + \bar{A} \cdot B \cdot C + A \cdot \bar{B} \cdot C + A \cdot B \cdot \bar{C}$$
$$f_2 = A \cdot B \cdot D + A \cdot B \cdot \bar{C} + A \cdot \bar{B} \cdot \bar{D} + \bar{B} \cdot C \cdot D$$

4.7. Show that a static hazard exists in the minimal network representing the function $\bar{A} \cdot C + B \cdot \bar{C}$. Devise minimal NOR and NAND equivalent networks, and investigate them for static hazards. If hazards exist, construct hazard-free networks.

References

1. MALEY, G. A. and J. EARLE, *The logical design of transistor digital computers,* Prentice-Hall.
2. ZISSOS, D. and G. W. COPPERWHITE, 'The design of minimal NOR/NAND logical circuits', *Electron. Engng.,* **37**, 9, 592–7, 1965
3. HUFFMAN, D. A., 'The design and use of hazard-free switching networks', *J. Assoc. Comp. Mach.,* **4**, 1, 1957

5. Flip-flops

In counting and *sequential networks* it is necessary to provide some form of memory element to record the state of the problem at any given time. The most commonly used device is known as the *flip-flop* or *bistable* element, which has two stable operating states. These correspond to the '0' and '1' logical levels, the output flipping from one stable state to the other upon demand. A flip-flop may therefore be defined as a device which stores binary information in the form of a '0' or a '1', and can be maintained indefinitely in either of the states. It can also be switched from one state to the other.

5.1 Types of memory element

Memory elements can broadly be divided into *non-volatile* and *volatile* types. Non-volatile memories retain their information almost indefinitely, even in the event of a power supply failure. Magnetic tapes, discs and cores are typical of this type. Volatile memories retain information only so long as the power supply to them is maintained. The majority of electronic memories are of the latter type, and are the subject of this chapter.

Two subgroups of volatile memories are *static memories* and *dynamic memories*. The information stored in a static memory is retained by the use of feedback between cross-connected gates and, subject to power supply continuity, the data stored is retained indefinitely. In dynamic memories, the data is stored in the form of an electrical charge on a capacitor; due to the inherent nature of capacitors, this charge decays with time, and periodically requires to be 'refreshed'. In this chapter the reader's attention is directed towards static memories. The operation of dynamic memories is best understood from the principles of semiconductor devices, and is covered at a later point in the book (see section 9.11).

5.2   Static memory elements

The three most popular types of *bistable memory elements* or *flip-flops* are
known as the *S-R* (Set-Reset), *J-K* and *D* (Delay or Data latch) elements, and are
described in the following sections. A further type, the *T* (Trigger or Toggle)
element is also widely used, and is constructed either from *J-K* or from
D elements.

The truth table describing the operation of the above memory elements
depends not only on the signals applied at a given instant of time, but also on
the history of the preceding events. That is, the truth table is 'dynamic' in
nature, and is subject to change.

5.3   The Set-Reset (*S-R*) flip-flop

The *S-R* flip-flop has two input lines, known as the *S*-line (Set line) and the
R-line (Reset line), and two output lines, namely the *Q*-output (the normal
output) and the $\bar{Q}$-output (or complementary output line). A typical block
diagram is shown in Fig. 5.1, and its truth table is given in Table 5.1. In this

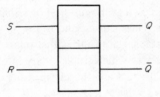

Fig. 5.1   The set-reset (*S-R*) flip-flop or bistable element

table, the column headed Q_n is the state of output *Q prior* to the application of
the conditions listed in the input columns. The column headed Q_{n+1} lists the
state of output *Q after* the application of the listed input conditions.

The 'storage' state of the *S-R* flip-flop occurs when the signal applied to both
the *S*- and *R*-lines is logic '0'. In this state the output is unchanged, and
$Q_{n+1} = Q_n$ (whatever the previous state of Q_n), illustrated in the first row of the
truth table. Output *Q* is 'reset' to zero by applying a logic '1' to the *R*-line ($S = 0$
at this time), shown in the second row of the truth table. The application of a
logic '1' signal to the *S*-line (when $R = 0$) causes output *Q* to be 'set' to the
logic '1' state.

When both the *S*- and *R*-line are energized by logic '1' signals, the state of the
output is not strictly defined since a simultaneous attempt is made both to set *Q*
to '1' and to reset it to '0'. What usually happens is that outputs *Q* and $\bar{Q}$ are no
longer complementary, and *both* output lines assume the same logic level;
depending on the type of gates used in the flip-flop, the output signals in this
state may either both be '0' or both be '1'. This operating state is usually
avoided, although there is no reason why it should not be used.

Two versions of *S-R* flip-flop are shown in Fig. 5.2. In the NOR version,

Table 5.1

Truth table of the S-R flip-flop

Inputs		Output		Comment	
S	R	Q_n	Q_{n+1}		
0	0	0 1	0 1	$Q_{n+1} = Q_n$	(Storage state)
0	1	0 1	0 0	$Q_{n+1} = 0$	(Reset)
1	0	0 1	1 1	$Q_{n+1} = 1$	(Set)
1	1	0 1	? ?	Not defined	

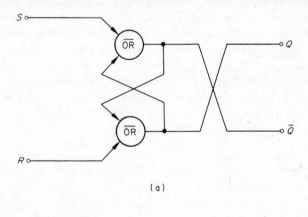

(a)

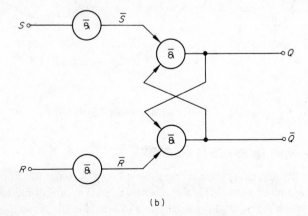

(b)

Fig. 5.2   Versions of the S-R flip-flop using (a) NOR gates, (b) NAND gates

Fig. 5.2(a), the S- and R-signals are applied directly to the principal gates, these gates being cross-connected to ensure that they form a self-latching system. This cross-connection provides positive feedback between the gates. In the NAND version, Fig. 5.2(b), the inputs are each applied to an invertor before being transmitted to the principle gates of the flip-flop; the reason is that logic '0' signals are required to activate NAND gates. Once again, the principal gates are cross connected to form a self-latching pair. Some versions of the NAND S-R flip-flop omit the input invertors, when the inputs are usually defined as $\bar{S}$ and $\bar{R}$.

5.4 Contact bounce elimination circuits

Electrical contacts are frequently used to provide input signals to logic systems. The contacts are frequently a source of electrical noise, which is due to 'contact bounce' arising from the mechanical properties of the contacts. If the 'noisy' signal is applied to a high speed counting system, the system 'sees' each noise impulse as a complete OFF-ON-OFF pulse; such a counting system would record the total number of noise 'pulses' (of which there may be several thousand in all) rather than record a single pulse. Two popular methods of eliminating contact bounce effects are shown in Fig. 5.3.

In Fig. 5.3(a), the first logic '1' signal applied to the S-input causes Q to become '1'. If the contacts of switch W 'bounce', the resulting train of pulses applied to the S-line have no further effect on the state of output Q. In Fig. 5.3(b), a logic '0' is used to control the switching operation of a NAND flip-flop.

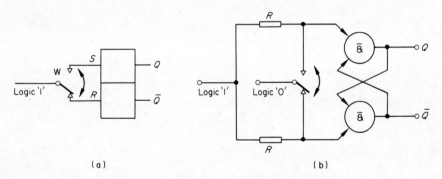

Fig. 5.3 Contact bounce elimination circuits

5.5 A clocked (gated) S-R flip-flop

In some circuits, the application of the S- and R-signals to the flip-flop is controlled by a *clock signal*. This signal is usually in the form of rectangular pulses, and is used to control the sequencing of events in the system. A typical clocked S-R flip-flop is shown in Fig. 5.4.

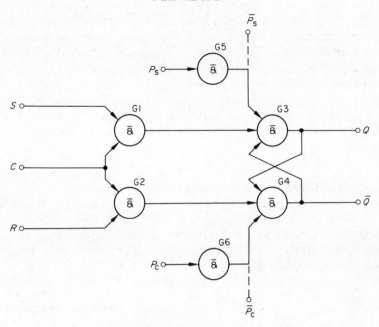

Fig. 5.4   A gated *S-R* flip-flop

When the clock signal, C, is logic '0' the operation of gates G1 and G2 is inhibited, and the signals on the S- and R-lines cannot be applied to the principal gates G3 and G4 of the flip-flop. When $C = 1$, gates G1 and G2 are opened to allow the signal on the input lines to be applied to the memory.

In certain circuits, it is desirable to be able to *preset* or *preclear* output Q by means of signals applied to additional control lines. In the circuit in Fig. 5.4, output Q can be preset to the logic '1' state by applying a logic '1' signal to the preset line, P_S, when $C = 0$. Similarly, Q is precleared (i.e., Q becomes '0') by applying a '1' signal to line P_C. In some circuits, gates G5 and G6 are eliminated, and presetting and preclearing are carried out by applying a '0' signal to lines $\bar{P}_S$ and $\bar{P}_C$, respectively.

5.6   Master-slave *S-R* flip-flop

In many high-speed counting circuits, race conditions (see chapter 7) are encountered which may result either in the output from the flip-flop being oscillatory or in it assuming an unpredictable state. To overcome this problem, a family of elements known as master-slave flip-flops were developed. The basic element is the *S-R* master-slave flip-flop, described below.

The principle of operation can best be understood by reference to Fig. 5.5. The sequencing of signals through the flip-flop is controlled by the clock signal, C, and when it has the logical value '0', switches G1 and G2 are open and G3 and G4 are closed. In this state, the data stored in the MASTER flip-flop, MFF, is

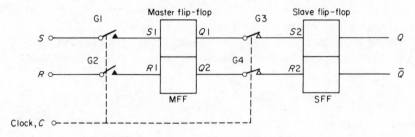

Fig. 5.5   The basis of the master-slave flip-flop

transmitted directly to the slave flip-flop, SFF, and thence to the output
terminals. In practice, switches G1 to G4, inclusive, are replaced by electronic
gates. When the clock signal changes to the logic '1' level, G1 and G2 close whilst
G3 and G4 open. In this operating state, new data is fed into MFF, whilst the
former state of MFF is stored in SFF, and continues to be presented at the
output terminals. A feature of the circuit described above is that data is
transferred through the flip-flop in a series of steps, and is related to the clock
pulse waveform (see Fig. 5.6) as follows:

1. *Clock signal* = 0 ($t < t_A$). In this state G1 and G2 are open, and G3 and G4
   are closed. Data stored in the master is transferred to the slave, and is
   presented at the output.
2. *Clock signal* = 1 ($t_B < t < t_C$). Here G1 and G2 are closed, and G3 and G4
   are open. New data is fed into the master; the data stored in the slave is
   unchanged.
3. *Clock signal* = 0 ($t > t_D$). G1 and G2 are open, and G3 and G4 are closed.
   The new data stored in the master is transferred to the slave, and is presented
   at the output.

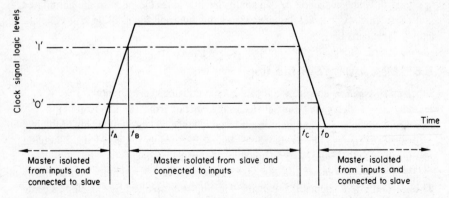

Fig. 5.6   Operating sequence of the master-slave flip-flop in terms of the clock pulse

Hence, *in the master-slave flip-flop, new data is applied to the master stage when the clock signal is '1', and is transmitted to the slave stage (i.e., to the output) when the clock signal is '0'.* To prevent electronic hazard conditions occurring, the sequencing of switches G1-G4 is arranged so that all four switches can never be simultaneously closed. In practice, this is brought about by the inherent delays in the electronic gates in the circuit.

A practical version of the master-slave *S-R* flip-flop is illustrated in Fig. 5.7, in which gates G1 to G4 correspond to switches G1 to G4, respectively, in Fig. 5.5. Synchronous operation of these gates is obtained by applying the clock signal to G1 and G2, and $\bar{C}$ to G3 and G4. In addition, gates G5 and G6 are incorporated to permit output Q either to be preset to '1' by signal P_S, or to be precleared (reset to '0') by signal P_C. In some cases, signal $\bar{P}_S$ may be applied to G1 and $\bar{P}_C$ to G2 in order to inhibit their operation during the presetting and preclearing cycle. The circuit shown can also accommodate several S-input lines and several R-input lines (shown by the dotted connections to gates G1 and G2), and this enables the flip-flop to perform logic functions in addition to its operation as a memory element.

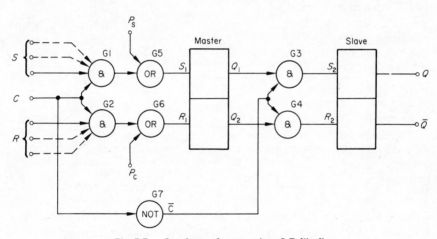

Fig. 5.7   One form of master-slave S-R flip-flop

5.7   The *J-K* master-slave flip-flop

Many early versions of flip-flop have been superceded by the master-slave *J-K* flip-flop, which is a modified version of the master-slave *S-R* flip-flop. A simplified block diagram of the *J-K* master-slave element is illustrated in Fig. 5.8. Gates G1, G2 and G7 in Fig. 5.8 correspond to gates with these numbers in Fig. 5.7, the essential differences between the two circuits is that, in Fig. 5.8, output Q is fed back to one input of G2 and output $\bar{Q}$ is fed back to one input

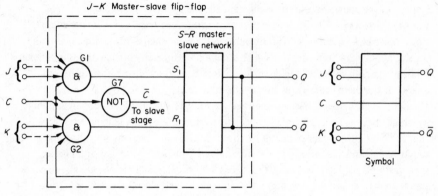

Fig. 5.8   A master-slave *J-K* flip-flop

of G1. The signal input to G1 is now designated the *J*-input, and the input to G2 is the *K*-input; the circuit can accommodate several *J* inputs and *K* inputs in the manner shown by the dotted links to G1 and G2 in Fig. 5.8. The truth table for the *J-K* flip-flop is given in Table 5.2.

Table 5.2

Truth table of the *J-K* master-slave flip-flop

Inputs		Output	Comment
J	*K*	Q_{n+1}	
0	0	Q_n	storage state
0	1	0	'reset'
1	0	1	'set'
1	1	$\bar{Q}_n$	'trigger' or 'toggle'

In the truth table, Q_n is the state of output Q prior to the application of the *n*th clock pulse, and Q_{n+1} is its state after the clock pulse has been applied.

When $J = K = 0$, both G1 and G2 are inhibited and the output is unchanged. When $J = 0$, $K = 1$ (row 2 of the truth table), a logic '1' is transmitted to the $\bar{Q}$ output of the master stage when the clock signal is at the logic '1' level, and is transferred to the $\bar{Q}$ output of the slave stage (i.e., Q is forced to be '0') when the clock signal falls to the '0' level. By a similar reasoning, it can be seen that when $J = 1$, $K = 0$, output Q becomes '1' after a clock pulse has been applied. Thus, the first three rows of the truth table of the *J-K* master-slave flip-flop correspond to those of the *S-R* flip-flop (with the *J*-input being equivalent to the *S*-input, and the *K*-input being equivalent to the *R*-input). The final row of Table 5.2 yields an interesting operating condition in which the output state of the flip-flop changes after the application of *each* clock pulse. This is known as *trigger operation* or *toggle operation*, and forms the basis of many counting circuits. This mode of operation arises from the fact that, when $J = K = 1$, the clock pulse causes the state of $\bar{Q}$ to be gated through the flip-flop to output Q, i.e., $Q_{n+1} = \bar{Q}_n$. When operated with $J = K = 1$, the device is described as a *T flip-flop*.

One form of master-slave *J-K* flip-flop using only NAND gates is illustrated in Fig. 5.9. In this circuit, preset and preclear inputs are provided, presetting being carried out by applying a logic '0' signal to input $\bar{P}_S$ and preclearing by applying a logic '0' to $\bar{P}_C$.

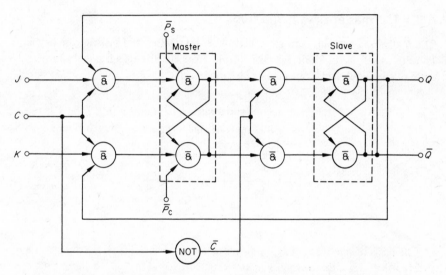

Fig. 5.9   One form of master-slave *J-K* flip-flop using only NAND gates

5.8   Functions performed by *J-K* flip-flops

The *J-K* master-slave flip-flop can be directly used to replace the S-R master-slave element provided that the condition $J = K = 1$ is not allowed to occur. When used as a *S-R* replacement, the *J*-inputs are used as *S*-inputs, and the *K*-inputs as *R*-inputs.

It may also be used as a trigger (T) flip-flop in either of the modes shown in Fig. 5.10. In diagram (a), both the *J*- and the *K*-lines are connected to logic '1' signals, the clock line being energized by incoming pulses (i.e., the clock line acts as the *T*-input line). In this mode of operation, the state of output Q changes state when the clock signal changes from '1' to '0'. A train of input pulses causes output Q to change in the sequence . . . 0, 1, 0, 1, 0 . . . . The same effect can be obtained in some logic families by leaving both the *J*- and *K*-input lines disconnected, as shown in Fig. 5.10(b); this connection should not be used in electrically noisy environments, since an induced noise signal on one of the input lines may cause the flip-flop to malfunction.

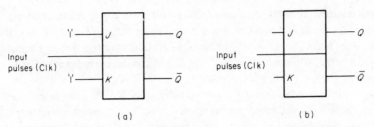

(a)   (b)

Fig. 5.10 (a) and (b)   The *J-K* flip-flop used as a trigger flip-flop

5.9   The master-slave *D* flip-flop

The *D* flip-flop has only a single control line, the *D*-line, and satisfies the logical conditions set out in Table 5.3. The output from the flip-flop is equal to the state of the input signal one clock pulse earlier.

Table 5.3

Truth table of the master-slave *D* flip-flop

Input *D*	Output Q_{n+1}
0	0
1	1

The basis of one form of *D* flip-flop is shown in Fig. 5.11(a). It consists of a *J-K* flip-flop (or, alternatively, a master-slave *S-R* flip-flop), whose input lines are energized by complementary input signals. The truth table of the *D* flip-flop is therefore equivalent to the second and third rows of the *J-K* flip-flop truth table (Table 5.2), with the *J*-input representing the *D*-input. It is also equivalent to the second and third rows of the truth table of the *S-R* flip-flop (Table 5.1), with the *S*-input replacing the *D*-input.

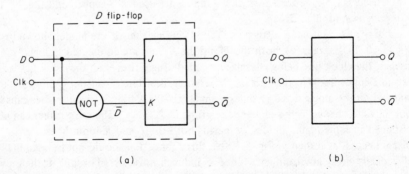

(a)   (b)

Fig. 5.11   (a) The basis of a *D* master-slave flip-flop, and (b) its circuit symbol

The D flip-flop is frequently used as a *data latch* element or *staticizer*, in which the state of one stage of a binary counter can be stored at the completion of a counting sequence. It is also used as a 'one-bit' delay element in serial arithmetic processors; it can be used, for example, to store the 'carry' or 'borrow' bit in an adder or subtractor, respectively.

The D flip-flop can also function as a trigger or toggle element by connecting it in the manner shown in Fig. 5.12. The input signal (the T-signal) is applied to the clock line. Resulting from this mode of operation, the D flip-flop can be used as a T element in counting circuits.

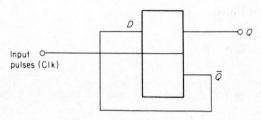

Fig. 5.12   Using the D flip-flop as a trigger or toggle element

5.10   Edge-triggered flip-flops

In an edge-triggered flip-flop, the output changes state in sympathy with the *leading edge* of the clock pulse (i.e., the '0' → '1' edge); this differs from the master-slave type, whose output changes in sympathy with the completion of the clock pulse. Truth tables of edge-triggered devices generally conform to those of equivalent master-slave types.

In these elements, the data signals (J, K or D) must be applied to the input(s) of the flip-flop for a minimum period of time known as the *set-up time* (typically 10 ns) prior to the clock pulse being applied. After the clock pulse has passed a threshold value (typically 1.5 V in TTL devices), the input information must be maintained for a period of time known as the *hold time* (typically a few nanoseconds). After the latter period of time, the input signals are 'locked-out' and have no further control over the state of the flip-flop output during the remainder of the clock waveform.

Edge triggering is brought about by modifying the logic circuitry of the flip-flop.

6. Asynchronous counting systems

Counting systems can broadly be divided into two categories, namely *asynchronous* and *synchronous* systems. Asynchronous counting systems function in a step-by-step manner, and a simple analogy of the process is given here. Suppose that we wish to add 893_{10} to 655_{10}; the simplest method is to add to the number stored in the counter (655_{10}) the number 1_{10} eight hundred and ninety three times. This sum illustrates the basic mechanics of asynchronous counting. The name asynchronous is derived from the fact that the states of the flip-flops do not change synchronously, since the '1' added at the least significant end of the sum causes a change to 'run through' or to 'ripple through' the stored value. As a result, these counters are sometimes known as *ripple-through* counters.

In synchronous counters, the states of all the flip-flops change synchronously under the control of a clock pulse. Counters of this type are described in chapter 7.

6.1 A pure binary counter

Table 6.1 illustrates the pure binary sequence for a four-bit counter, in which D is the least significant bit. From this table it is possible to deduce a simple rule for counters which operate in the 'forward' mode, i.e., for increasing value of count. From the table, it is observed that *each bit, except the least significant bit, changes state when the next less significant bit changes from '1' to '0'. The least significant bit changes state following the application of each input pulse.* These conditions are satisfied by the T flip-flop described in chapter 5.

A basic form of asynchronous pure binary counter is illustrated in Fig. 6.1(a). This circuit uses four J-K flip-flops, each connected to operate as a T flip-flop. The preclear (P_C) input lines or reset lines are connected to a common wire; a logic '1' applied to this line causes all outputs to be reset to zero.

Table 6.1

Pure binary code

Decimal value		A	B	C	D	
0		0	0	0	0	
1		0	0	0	1	
2		0	0	1	0	
3		0	0	1	1	
4		0	1	0	0	
5		0	1	0	1	
6	Forward count	0	1	1	0	Reverse count
7		0	1	1	1	
8		1	0	0	0	
9		1	0	0	1	
10		1	0	1	0	
11		1	0	1	1	
12		1	1	0	0	
13		1	1	0	1	
14		1	1	1	0	
15		1	1	1	1	

The input pulses arrive in a random sequence, and the pulse width may vary without affecting the accuracy of the count in any way. Flip-flop FFD changes state at every $1 \rightarrow 0$ edge, i.e., the trailing edge, of the input pulse. FFC changes state every time the output of FFD changes from '1' to '0'. Ideally both flip-flop outputs change simultaneously, but, for reasons given below, this does not occur. As the forward count proceeds, the outputs of the flip-flops change in accordance with Table 6.1, to the count of 15_{10}, when the outputs of all the flip-flops are in the '1' state. The next input pulse resets FFD to zero, and this change ripples through the counter making all the outputs zero.

In practice each flip-flop introduces a propagation delay, and there is a time delay between the change in the input signal occurring, and the corresponding output change taking place. This is assumed to have a value τ in each of the flip-flops here. Figure 6.1(b) illustrates the state of affairs during the transition from the count of 3_{10} to 4_{10}. The output of FFD changes state τ s after the $1 \rightarrow 0$ edge of its input pulse. Thus τ s after the end of the fourth pulse the output of FFD falls to zero, giving a transient count of $0010_2 = 2_{10}$. After a further period of τ s, the output of FFC becomes zero, giving a second transient count of 0000. This change triggers FFB into the '1' state after a further delay of τ, giving a final count of $0100_2 = 4_{10}$. In the T flip-flop the $0 \rightarrow 1$ transition does not alter the output state, and the output of FFA remains stable at '0'. The change from 3_{10} to 4_{10} is thus achieved in 3τ s, with two erroneous states having occurred during the transition. Normally τ is small, being a fraction of a microsecond or less, in electronic devices, and a few milliseconds in pneumatic devices. In many applications it may be ignored. It is in high speed systems that the propagation delay is significant.

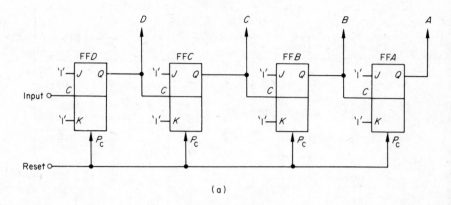

(a)

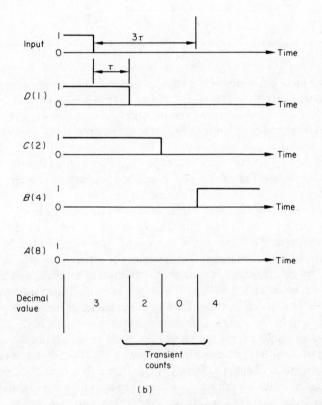

(b)

Fig. 6.1   (a) A ripple-through pure binary counter. The effect of the individual time delays in the transition from 3_{10} to 4_{10} are shown in (b)

In the case of the 16th pulse, each flip-flop must change its output state, and the final state (0000) is achieved after 4τ s.

The counter can be preset to some initial value by generating a preset signal at the final stage of the count. This signal is then applied to appropriate inputs of the flip-flops. If the initial condition is to be $ABCD = 0011$, a '1' signal must be applied to the P_S-lines of flip-flops C and D, and to the P_C-lines of flip-flops A and B.

6.2 Reversible binary counting

A rule for reverse binary counting can be deduced by reading Table 6.1 in the reverse order, and is stated as follows.

Each stage, except the least significant stage, changes state when the less significant stage changes from '0' to '1'. The least significant stage changes state for each input pulse.

Thus if the state of the counter at the commencement of the reverse count is 0001 (1_{10}), then the next pulse must set it to 0000, and following successive pulses to 1111 (15_{10}), 1110 (14_{10}), etc. Using T flip-flops, reverse counting is achieved by connecting the $\bar{Q}$ output of the less significant stage to the C-line of the following flip-flop, as shown in Fig. 6.2. The operation can be followed by recalling that the output from a T flip-flop changes state whenever its input changes from '1' to '0'.

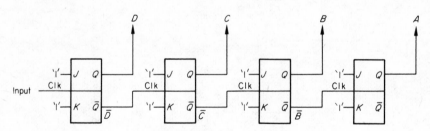

Fig. 6.2 A reverse or 'down' counter using T flip-flops

Reverse counting can also be achieved in the forward counter, shown in Fig. 6.1(a), by monitoring the outputs from $\bar{A}, \bar{B}, \bar{C}$, and $\bar{D}$. An inspection of Table 6.1 shows this to be correct, since, if each term in the group is complemented, the result is the 15's complement of the number.

6.3 Bidirectional counters

Bidirectional counters can be developed by combining the simple forward and reverse counters, by AND gating the output signals with a control signal, which has the value '1' or '0'. This control signal then allows either the forward count, or the reverse count to proceed.

An alternative rule for *forward counting* is deduced by inspecting Table 6.1. It is observed that *each stage, except the least significant stage, changes state following the condition when all the less significant stages are in the '1' state. When counting in reverse, each stage, except the least significant stage, changes state following the condition when all the less significant stages are in the '0' state. When counting in both forward and reverse the least significant stage changes state following each input pulse.* A network which satisfies these conditions is shown in Fig. 6.3.

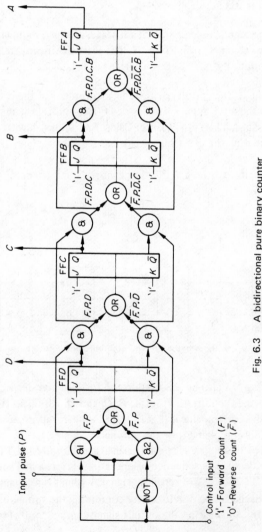

Fig. 6.3   A bidirectional pure binary counter

A '1' applied to the control input activates AND 1 and inhibits AND 2. Since the output of AND 2 is zero, all the lower AND gates are inhibited. A series of pulses applied at input P result in a forward, or 'up', count taking place, since a signal is only applied to the flip-flop when the input pulse is present AND all the less significant stages are in the '1' state. When the input pulse falls to zero, the appropriate outputs change state. When the control signal is '0', the upper AND gates are inhibited, and AND 2 is activated. A signal is applied to each stage only when the input pulse is present AND the outputs of the less significant stages are in the '0' state. This results in a reverse, or 'down', count.

6.4 Non-pure binary counters

All the outputs of the flip-flops, in the counters described earlier, are zero simultaneously only on a count of zero, or after a count of 2^N, where N is the number of stages in the counter, i.e., with four stages the count is zero at the count of zero, and again after the 16th pulse. If the counter is to be reset to zero after some other value, e.g., 5, 6, 10, etc., additional logic circuits are required. Many circuits are possible, a popular counter circuit being discussed here.

AN 8421 BCD COUNTER

A circuit which generates the 8421 BCD code (see also section 1.7) is shown in Fig. 6.4. The counter uses J-K flip-flops connected to function as T elements; in order to simplify the diagram, the J- and the K-input lines are omitted, and these

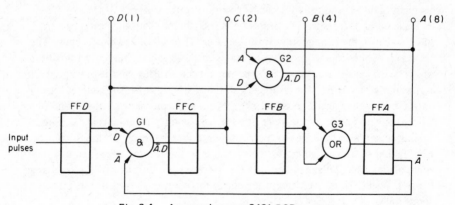

Fig. 6.4   An asynchronous 8421 BCD counter

would be connected to a logic '1' signal (or left open-circuited in TTL gates in a noise-free environment). The sequence of operations is given in Table 6.2.

Initially, when all outputs are zero, the signal $\bar{A}$ which is fed back to G1 opens this gate and connects the four flip-flops together in the form of a basic pure binary counter. At the same time, G2 is open (since $A = 0$) and has no

Table 6.2

Sequence of events in an 8421 BCD counter

Pulse number	State of outputs	State of gates	
	ABCD	G1	G2
Initial conditions	0000		
1	0001		
2	0010		
3	0011		
4	0100	open	closed
5	0101		
6	0110		
7	0111		
8	1000		
9	1001	closed	open
10	0000	open	closed

effect on the operation of the counter in its early stages of operation. During the time interval when the first seven pulses are applied to the counter, it functions normally, i.e., as described for the 'forward' counter in Fig. 6.1(a).

Pulse number 8 results in outputs *ABCD* changing from the combination 0111 to 1000; the change in output *A* causes G1 to close and G2 to open. The net result of the change in operating states of these gates is that any transition which occurs at the output of FF*D* is transmitted to FF*A*, and is prevented from being applied to FF*B*. As a result, the states of FF*C* and FF*B* remain at logic '0' until G1 is opened again. Pulse number 9 causes output *D* to change from '0' to '1'; this has no effect on the state of FF*A*, whose output remains at '1'. The trailing edge ('1' → '0' edge) of input pulse number 10 results in the output of FF*D* falling to '0' and, since this corresponds to the trailing edge (i.e., a '1' → '0' edge) of the clock pulse applied to FF*A*, it causes the output of FF*A* to change from '1' to '0'. Thus, after ten input pulses the counter recycles to its initial condition of $A = B = C = D = 0$.

Problems

6.1. Show how four bistable devices may be used in a binary counter. Sketch waveform diagrams for each stage to illustrate the operation of the counter.

6.2. Design a ripple-through counter which works in the following code, and state the 'weight' of each digit.

0000
0001
0011
0100
0101
0111
1100
1101
1111
———
0000
etc.

6.3. Design a counter which works in the following code, and state the 'weights' of each digit.

0000
0001
0010
0011
0111
1000
1001
1010
1011
1111
———
0000
etc.

7. Synchronous counting systems

In asynchronous systems the state of the system is independent of the clock or input pulse since it takes a finite time for all the changes to 'ripple through' the system. The time taken to count a pulse in an asynchronous system is dependent on the total number of stages involved. In *synchronous* systems the clock or input pulse initiates all the changes simultaneously, and the total time taken to count one pulse is generally much less than that of an equivalent asynchronous counter.

Synchronous systems are usually more complex than asynchronous systems, since it is necessary to prepare the input logic gates to count the pulse, in advance of the pulse being received. This complexity is often offset by the increase in operating speed.

Synchronous systems are also prone to stability problems which do not normally exist in asynchronous systems. These problems are discussed more fully in sections 7.1 and 7.2.

7.1 Stability

Unstable operating conditions can arise in logical networks as a result of changes within the network. Consider the network in Fig. 7.1. If $x = 0$ the AND gate is inhibited giving an output $y = 0$, which is a stable operating state. When $y = 0$ and x is changed to '1' both inputs to the AND gate are activated ($\bar{y} = 1$), resulting in y rising to the logical '1' level. This signal is fed back via the NOT gate to its own input, giving a '0' at the lower AND input. This inhibits the AND gate and y falls to the logical '0' level. When the '0' is fed back, y rises again to the '1' level. This process is repeated indefinitely so long as $x = 1$.

In this case the output of the network is oscillatory, the periodic time of the oscillations being dependent on the time taken for the logic signal to propagate through the network. By inserting a time delay at point A, in Fig. 7.1, the frequency of the oscillations can be controlled.

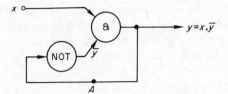

x o

& ────── $y = x . \bar{y}$

NOT $\bar{y}$

A

Fig. 7.1   Instability in a simple network

The above argument is based on intuitive reasoning which is satisfactory for simple systems, but a more analytical approach is required in complex systems. A detailed discussion is beyond the scope of this book, but sufficient information is given here to outline the principles involved. Further information can be obtained from the Refs. 1 and 2.

One method of displaying the variations in output is the *Y-map* or flow table. The *Y*-map for the function $y = x . \bar{y}$, generated in Fig. 7.1, is shown in Fig. 7.2. Horizontal movements in the map correspond to changes in the input variable x, and vertical movement to changes in the output y. Cross-hatched areas represent stable operating states. The entries outside the *Y*-map are known as the *energization* states, and those within the map are known as the *operation* states. The energization states define the operation states that the system must assume. The resulting operation states become the energization states for the following step. Stable operation is achieved when the operation states lead to their own energization states, i.e., when the energization and operation states are identical. Only one such state occurs in Fig. 7.2, that when $x = 0$, $y = 0$, which is cross-hatched.

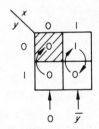

Fig. 7.2   The Y-map of the network in Fig. 7.1

The operation states are deduced from the logical equations of the system and the given input state. Thus for $y = x . \bar{y}$

(a) when $x = 0$ then $y = 0 . \bar{y} = 0$
(b) when $x = 1$,   then $y = 1 . \bar{y} = \bar{y}$.

The y-values of the function are '0' for $x = 0$, and are plotted in the upper and lower left-hand cells. When $x = 1$ the operation state is always $\bar{y}$, giving the values in the right-hand cells which are opposite to the y-value outside the cell.

The operation of the network may be analysed as follows. Initially if $y = 1$ and $x = 0$, the operation state is '0'. This becomes the new energization state. Corresponding to this energization state, the operation state is '0' and the system moves to the cell $x, y = 0,0$. This is a stable operating state since the energization and operation states are equal in value, and the output locks in this state. If x is now changed to '1' the operation state becomes '1', and the state of the system moves to the cell defined by the energization states $x, y = 1,1$. This results in a new operation state of $y = 0$. The net result is that the output oscillates between the '0' and '1' levels, as shown by the arrows in the two right-hand cells, the periodic time of oscillation depending on the network delays. When x is changed to '0' again, the network always returns to its stable operating state $x, y = 0,0$.

A succession of two or more unstable states, one leading to another is known as a *cycle*. In nearly all cases in logical networks a cycle results in the system malfunctioning if it is allowed to occur.

7.2 Race conditions

Most logic systems have several input lines, and if the change in output commanded is not a unit-distance change, then the actual mechanics of the change may be undefined. Suppose the system has two outputs, y_1 and y_2, which are to change from the state 0,0 to 1,1. Due to the physical nature of logic devices, one will operate more rapidly than the other, and the output will change in unit-distance fashion as follows

$$0,0; \ 0,1; \ 1,1 \quad \text{or} \quad 0,0; \ 1,0; \ 1,1.$$

Consider the NOR S-R flip-flop circuit in Fig. 7.3(a). The outputs from the NOR gates are

$$y_1 = \bar{x}_1 \cdot \bar{y}_2$$
$$y_2 = \bar{x}_2 \cdot \bar{y}_1.$$

Clearly, when

$$
\begin{array}{ll}
x_1 = 0, & y_1 = \bar{y}_2 \\
x_1 = 1, & y_1 = 0 \\
x_2 = 0, & y_2 = \bar{y}_1 \\
x_2 = 1, & y_2 = 0.
\end{array}
$$

These conditions are tabulated at the foot of the Y-map in Fig. 7.3(b) together with the appropriate operation states on the map.

It is assumed that the network is initially in the stable state $y_1, y_2 = 0,1$, with $x_1, x_2 = 0,0$. If the input states are now changed to $x_1, x_2 = 0,1$, the output changes in the following manner, shown by arrows on the diagram.

$$y_1, y_2 = 0,1; \ 0,0; \ 1,0; \ 1,0 \quad \text{(stable)}.$$

The first change $0,1 \rightarrow 0,0$ corresponds to a change of the state of y_2 since one input to NOR 2 is '1'. The changed value of y_2 alters the input conditions to

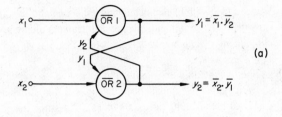

(a)

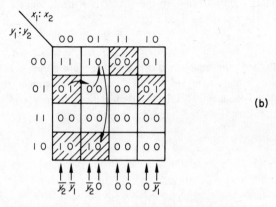

(b)

Fig. 7.3 (a) A NOR memory element, and (b) the Y-map illustrating the changes that take place when input x_2 changes from '0' to '1'

NOR 1 $(x_1, y_2 = 0,0)$, leading to the new output conditions $y_1, y_2 = 1,0$. It should be noted that all the changes are unit-distance. A change of input conditions back to $x_1, x_2 = 0,0$ results in a horizontal movement on the Y-map to the stable state $y_1, y_2 = 1,0$, i.e., no change in output.

Had the input conditions changed from $x_1, x_2 = 0,1$ to $x_1, x_2 = 1,1$ the final stable output condition would by $y_1, y_2 = 0,0$, i.e., both outputs are zero. If both inputs are reduced to zero simultaneously, the outputs are commanded to change from 0,0 to 1,1 as shown in Fig. 7.4. As stated above, one of the gates

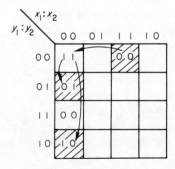

Fig. 7.4 A critical race condition

would operate more rapidly than the other and either of the two intermediate values 0,1 or 1,0 would first occur before the 1,1 state, as illustrated in Fig. 7.4. Since both lead to stable operating states, the condition $y_1,y_2 = 1,1$ is never achieved.

If more than one operation state changes at a time, a *race condition* is said to exist. If the race terminates in two or more unequal states, it is said to be a *critical race*. A critical race condition exists in Fig. 7.4 since the race terminates in two unequal stable conditions. In the NAND *S-R* flip-flop of Fig. 5.2(b) similar critical race conditions hold.

In some networks two or more patterns exist which lead to the same stable state. Such a race is described as *noncritical*. An example is given in Fig. 7.5. It is assumed initially that the network is in the stable state $y_1,y_2 = 0,0$, when the inputs x_1,x_2 are changed from 0,1 to 0,0. The new operation state is 1,1, and the output changes from 0,0 to 1,1 by either of the patterns 0,0; 0,1; 1,1, or 0,0; 1,0; 1,1. In either case the same stable state is finally reached.

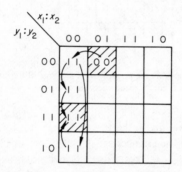

Fig. 7.5   Non-critical race conditions

Combinations of cycles sometimes occur if information is fed back, an example being given in Fig. 7.6. The *Y*-map for the network in Fig. 7.6(a) is given in Fig. 7.6(b). Assume that the network is in the stable state $y_1,y_2 = 0,1$ with input signals $x_1,x_2 = 1,0$. When the inputs are changed to 1,1, the y_1 output remains at '0' and the y_2 output oscillates, giving the cycle $y_1,y_2 = 0,0$; 0,1; 0,0; 0,1, etc. If both inputs are simultaneously reduced to zero, when $y_1,y_2 = 0,1$, a critical race condition occurs and the final state of the output could be 0,1 or 1,0.

Where feedback is applied around flip-flops, care should be taken to ensure that cycles and critical races do not occur. Non-critical races can be allowed to exist if they do not affect the performance of the system. One method of overcoming these problems, where flip-flops are employed, is to use master-slave flip-flops (chapter 5) in which the inputs and outputs of the flip-flop are never connected directed together.

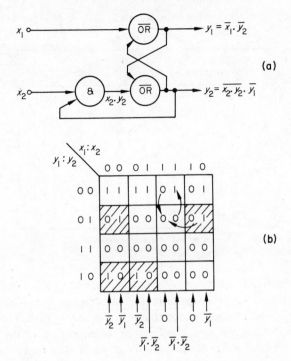

Fig. 7.6   The *Y*-map in (b) illustrates how unstable operation results by the application of a feedback loop in (a) to a NOR *S-R* flip-flop

7.3   The design of parallel counters

Several methods of designing parallel counters have been evolved including analytical[3] and map[4,5] methods. The main advantage of map methods are that they are simple to use and understand. Their principle disadvantage is that they are difficult to deal with when more than four to six variables are involved.

An example using the map method follows. Consider a counter which is to count in 2421 BCD, and is to use *J-K* flip-flops. The truth table for the code is given in Table 7.1, where *A* is the most significant digit and *X* signifies a 'can't happen' condition. The decimal values are marked on the Karnaugh map in Fig. 7.7.

In parallel counters the outputs of all the flip-flops change simultaneously. To do this the input logic must be prepared in advance of the incoming pulse to be counted. Immediately after the *n*th pulse has passed, the input logic must be prepared for the $(n + 1)$th pulse, and must be held in that state until the pulse is received. After this the logic must be prepared for the $(n + 2)$th pulse. It is therefore necessary in any given state, to study the following state to be assumed by the counter.

Table 7.1

2421 BCD code

Decimal count	A	B	C	D
0	0	0	0	0
1	0	0	0	1
2	0	0	1	0
3	0	0	1	1
4	0	1	0	0
5	0	1	0	1
6	0	1	1	0
7	0	1	1	1
8	1	1	1	0
9	1	1	1	1
10–15	X	X	X	X

X = 'can't happen' condition

Consider flip-flop C. At the zero count its output is '0', and it must be retained at this level for the count of unity, after which it must change to the '1' level. This level is maintained for the counts of 2_{10} and 3_{10}, after which it must change back to '0' and remain at this level for the count of 5_{10}. The logical level must rise to '1' on the 6th pulse, and remain at this level to the count of 9_{10}. On the 10th pulse its output becomes zero and the cycle begins again. In order to reproduce this cycle accurately the input conditions to the flip-flop for a given change in output must be known. These changes for the J-K flip-flop are listed in Table 7.2. The code change information from Table 7.1, is transferred to a code change map for each flip-flop, and thence to the input logic maps using Table 7.2.

This procedure is illustrated in Fig. 7.8 for flip-flop A. From Table 7.1 it is observed that the flip-flop output is zero from the count of zero to 6_{10}. In the code change map in Fig. 7.8 the cells equivalent to 0_{10} to 6_{10}, in Fig. 7.7, are

CD \ AB	0 0	0 1	1 1	1 0
0 0	0	4	X	X
0 1	1	5	X	X
1 1	3	7	9	X
1 0	2	6	8	X

Fig. 7.7　　The Karnaugh map for the 2421 BCD code in Table 7.1

Table 7.2

Input conditions to the *J-K* flip-flop

	Input logic levels	
Output change	*J*	*K*
0 → 1 (change from '0' to '1')	1	*X*
1 → 0 (change from '1' to '0')	*X*	1
0 → 0 (maintain '0')	0	*X*
1 → 1 (maintain '1')	*X*	0

X = 'don't care', i.e., could be '0' or '1'

marked $0 \rightarrow 0$. After the 7th pulse the output of flip-flop A must change to the '1' level. This is recorded in the cell in the *code change map* equivalent to 7_{10} as a $0 \rightarrow 1$ change. This state is maintained after the 8th pulse, and is shown as a $1 \rightarrow 1$ change in the 8_{10} cell of the code change map. The final change after the 9th pulse is shown as a $1 \rightarrow 0$ change in the 9_{10} cell of the code change map.

By referring to Table 7.2, the input logic maps can now be completed. The J_A and K_A maps in Fig. 7.8 refer to the J and K input lines, respectively, of the 'A' flip-flop. In cells in the J_A and K_A maps corresponding to those marked $0 \rightarrow 0$ in the code change map, 0's and X's are marked, respectively. This follows from the third row of Table 7.2. In the J_A and K_A cells corresponding to the cell marked $0 \rightarrow 1$ in the code change map, '1' and 'X' are marked, respectively. The $1 \rightarrow 1$ and $1 \rightarrow 0$ cells are then dealt with giving the final input logic maps. When all the changes have been covered the input logic is minimized, giving

$$J_A = B.C.D \quad \text{and} \quad K_A = D.$$

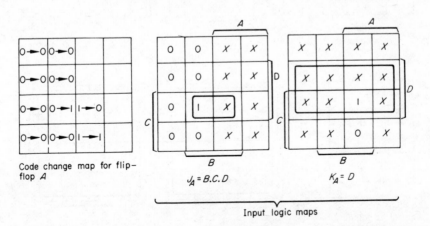

Code change map for flip-flop A

$J_A = B.C.D$

$K_A = D$

Input logic maps

Fig. 7.8 The input logic maps for any counter can be deduced by writing down the code change information on the Karnaugh map and then grouping terms on the input logic maps, illustrated here for flip-flop A

Maps for flip-flops B, C, and D are given in Fig. 7.9. The input equations are given below, and the corresponding logic block diagram, using master-slave flip-flops, is given in Fig. 7.10.

$$J_B = C \cdot D \qquad K_B = A \cdot D$$
$$J_C = D \qquad K_C = A \cdot D + \bar{B} \cdot D = D \cdot (A + \bar{B})$$
$$J_D = 1 \qquad K_D = 1$$

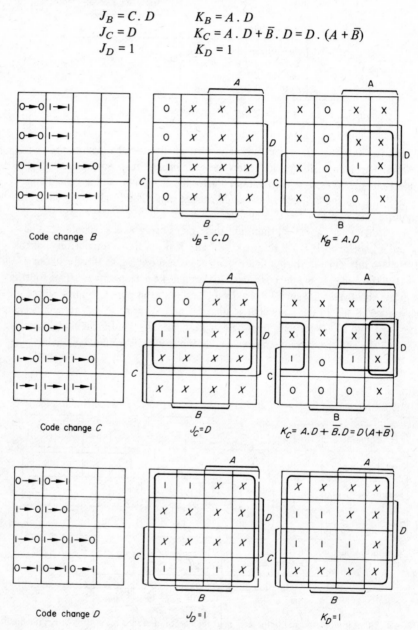

Fig. 7.9   Input logic maps for flip-flops B, C and D

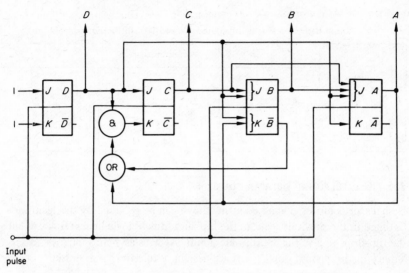

Fig. 7.10 A 2421 BCD synchronous counter using *J-K* flip-flops

Alternative modes of input connection are shown in Fig. 7.10, since the logic AND function can be performed internally in some *J-K* flip-flops. For flip-flop *C* the AND function on the *K* inputs is performed externally, while in flip-flops *A* and *B* the internal AND function is used.

Designs may be carried out using *S-R* and *T* flip-flops using this technique. It is essential to use master-slave versions in parallel counters, otherwise instability and critical race conditions can occur when feedback is applied around the flip-flop, as illustrated in Fig. 7.6.

Input logic levels are listed in Table 7.3 for *S-R*, and *T* flip-flops for given output changes. The logic required at the inputs to the flip-flops, for the 2421 BCD counters, is given in Table 7.4. In the case of *T* flip-flops, no input logic is required for flip-flop *D* since the output changes state at each incoming pulse.

Table 7.3
Input conditions to *S-R* and *T* flip-flops for given output changes

Output change	Input logic levels		
	S	R	T
$0 \to 1$	1	0	0
$1 \to 0$	0	1	0
$0 \to 0$	0	X	1
$1 \to 1$	X	0	1

X = 'don't care'

Table 7.4

Input logic to S-R and T flip-flops for parallel 2421 BCD counters

Flip-flop	Input logic		
	S	R	T
A	$\bar{A} . B . C . D$	D	$\bar{B} + \bar{C} + \bar{D}$
B	$\bar{B} . C . D$	$A . D$	$\bar{C} + \bar{D} + B . \bar{A}$
C	$\bar{C} . D$	$A . D + \bar{B} . C . D$	$\bar{D} + \bar{A} . B . C$
D	$\bar{D}$	D	Input pulse

7.4   Bidirectional parallel counters

Designs for counting up and counting down can be producd by the technique outlined in the previous section, the 'up' and 'down' logic being OR-gated into the flip-flops to give the appropriate count. As with asynchronous systems, both 'count-up' and 'count-down' lines are necessary to ensure that only the correct logic is activated.

Consider a counter which is to count up and down in 2421 BCD. The 'up' count design is outlined in section 7.3, the 'down' count design being considered here. The code is tabulated in Table 7.5 in the reverse counting order. The 'down' count proceeds from a higher number to a lower one, so that when the counter is, say, in the 6_{10} state the logic for the 5_{10} state has to be prepared. The code change maps and input logic maps are shown in Fig. 7.11 for the code in Table 7.5.

Table 7.5

Truth table for 2421 BCD 'down' count

Decimal count	State of flip-flop			
	A	B	C	D
9	1	1	1	1
8	1	1	1	0
7	0	1	1	1
6	0	1	1	0
5	0	1	0	1
4	0	1	0	0
3	0	0	1	1
2	0	0	1	0
1	0	0	0	1
0	0	0	0	0

The input logic to the bidirectional counter is obtained by combining the input logic for the up count, from Figs. 7.8 and 7.9, with the down count logic, together with suitable excitation lines to ensure that only the correct gates are energized. In the following the 'up' count line, U, is energized ('1') when the

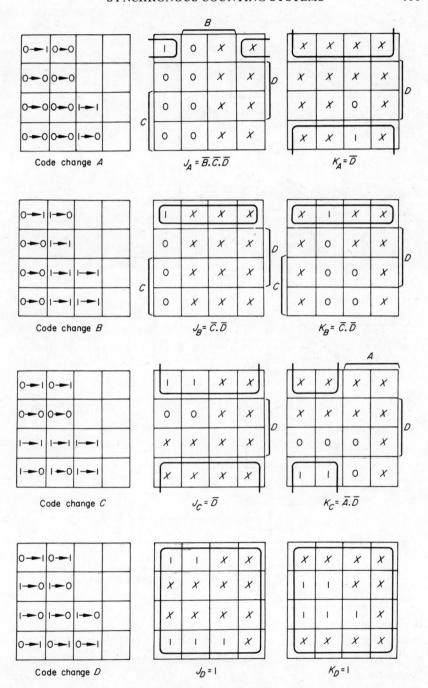

Fig. 7.11　Code change and input logic maps for a counter which counts 'down' in 2421 BCD using *J-K* flip-flops

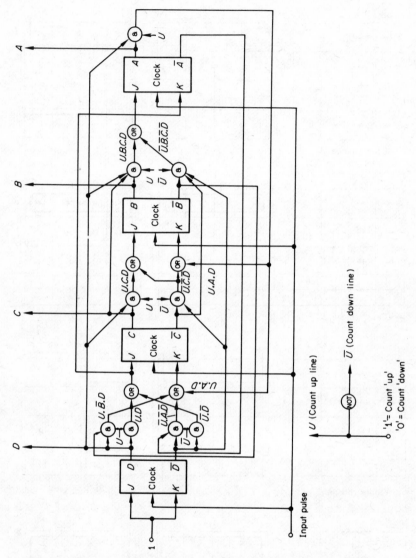

Fig. 7.12   Bidirectional 2421 BCD counter using *J-K* flip-flops

count proceeds from zero to nine, and the down count line, $\bar{U}$, is energized when the count proceeds from nine to zero. The logic block diagram of the counter is given in Fig. 7.12.

$$J_A = U \cdot B \cdot C \cdot D + \bar{U} \cdot \bar{B} \cdot \bar{C} \cdot \bar{D}$$
$$K_A = U \cdot D + \bar{U} \cdot \bar{D}$$
$$J_B = U \cdot C \cdot D + \bar{U} \cdot \bar{C} \cdot \bar{D}$$
$$K_B = U \cdot A \cdot D + \bar{U} \cdot \bar{C} \cdot \bar{D}$$

$$J_C = U \cdot D + \bar{U} \cdot \bar{D} \quad (= K_A)$$
$$K_C = U(A \cdot D + \bar{B} \cdot D) + \bar{U} \cdot \bar{A} \cdot \bar{D}$$
$$J_D = U \cdot 1 + \bar{U} \cdot 1 = 1$$
$$K_D = U \cdot 1 + \bar{U} \cdot 1 = 1$$

The general design principle may be extended to deal with any given code. The input logic for four codes, using J-K master-slave flip-flops, is given in Table 7.6. Pure binary counters with N stages can readily be constructed with J-K flip-flops, since the J and K inputs for the Mth stage are both $A \cdot B \cdot C \cdot \ldots \cdot J \cdot K \cdot L$ for the 'up' count and $\bar{A} \cdot \bar{B} \cdot \bar{C} \cdot \ldots \cdot \bar{J} \cdot \bar{K} \cdot \bar{L}$ for the 'down' count.

Counters which count up in one code and down in another, can be constructed using the techniques outlined here.

Table 7.6
Input logic for several counters using J-K flip-flops

Input line	Code							
	8421 BCD		5211		642(−3)		8421 Pure binary	
	Up	Down	Up	Down	Up	Down	Up	Down
J_A	$B \cdot C \cdot D$	$\bar{B} \cdot \bar{C} \cdot \bar{D}$	$B \cdot C$	$\bar{D}$	$C + B \cdot D$	$\bar{D}$	$B \cdot C \cdot D$	$\bar{B} \cdot \bar{C} \cdot \bar{D}$
K_A	D	$\bar{D}$	$B \cdot C$	$\bar{D}$	$\bar{B} \cdot D + C \cdot D$	$\bar{C} + \bar{B} \cdot D$	$B \cdot C \cdot D$	$\bar{B} \cdot \bar{C} \cdot \bar{D}$
J_B	$C \cdot D$	$A \cdot \bar{D}$	C	$\bar{D}$	$A + \bar{C}$	$C + \bar{D}$	$C \cdot D$	$\bar{C} \cdot \bar{D}$
K_B	$C \cdot D$	$\bar{C} \cdot \bar{D}$	C	$\bar{C}$	$A + \bar{C}$	$\bar{A} + D$	$C \cdot D$	$\bar{C} \cdot \bar{D}$
J_C	$\bar{A} \cdot D$	$A \cdot \bar{D} + B \cdot \bar{D}$	D	$B + \bar{D}$	B	$A + \bar{B}$	D	$\bar{D}$
K_C	D	$C \cdot \bar{D}$	1	1	$\bar{A} + B$	$\bar{B} + \bar{D}$	D	$\bar{D}$
J_D	1	1	1	1	1	1	1	1
K_D	1	1	$B \cdot C$	$\bar{B} \cdot \bar{C}$	1	1	1	1

7.5 Shift registers

A shift register comprises a number of cascaded flip-flops which contain a pre-arranged pattern of 1's and 0's. The application of a pulse, known as the *shift pulse*, moves the pattern by one step along the register.

Imagine four flip-flops A, B, C, D in a shift register, D being the least significant, all set to the '1' state when a series of shift pulses are applied. The result is shown in Table 7.7. After four shift pulses the complete pattern is moved along, and out of the register, the 1's being replaced by 0's. In general, if there are N flip-flops, it requires N shift pulses to shift all the data out of the register.

Master-slave J-K elements and D flip-flops are frequently used in these applications. The basis of operation of two stages of a shift register is illustrated in Fig. 7.13. The initial values stored at outputs C and D are '1', a '0' signal is applied to the J-input of FFD, and a '1' signal is applied to its K-input. *In all normal types of shift register, complementary logical signals must be applied to the J- and K-inputs.* The reason for this can be understood from the description below. The conditions mentioned above correspond to the initial conditions listed in Table 7.7 for the flip-flops C and D. The operation of the circuit during the first two clock pulse cycles is now described. At time t_1 (see Fig. 7.13(b)), when the shift signal (clock signal) is '0', the master stages of both flip-flops are isolated from their respective input signals. At t_2, the shift signal changes to logic '1' which results in the master stages (a) being isolated from the slave stages, and (b) being connected to their respective input lines. This causes the Q-output of the *master stage* of FFD to store a '0', and the $\bar{Q}$-output of this stage to store a '1'. Since the master and slave stages are isolated from one another, the Q-output of FFD continues to store a '1'. At this time, the master stage of FFC is connected to the slave stage of FFD, and therefore stores logic '1'; there is, therefore, no change in the condition of FFC. Hence *the data applied to the input lines (which must be complementary) of master-slave flip-flops is shifted into the master stages when the shift signal changes from '0' to '1'.*

Table 7.7

Movement of data in a shift register

Shift pulse number	Direction of movement of data				Input data
	⟵				
	State of flip-flops				
	A	B	C	D	
Initial state	1	1	1	1	0
1	1	1	1	0	0
2	1	1	0	0	0
3	1	0	0	0	0
4	0	0	0	0	

The above conditions are maintained between t_2 and t_3, at which time the shift signal falls to logic '0'. This causes the master stages (1) to be isolated from the input lines, and (2) to be connected to the slave stages. Consequently, *when the shift pulse signal changes from '1' to '0', the data in each master stage is transferred to its slave stage.* As a result of the above operation, the '0' stored in the Q-output of the master stage of FFD is transferred to its slave stage, and output D changes from '1' to '0'. At the same time, the state of the master stage of FFC is shifted into its slave stage, but since both store 1's, there is no apparent change in the output of FFC at time t_3.

During the second shift pulse, input data is shifted into both master stages at t_4 (0's in both cases), and is transferred to their slave stages at t_5. Hence, at time t_5, output D remains unaltered at logic '0', and output C changes from '1' to '0'. Readers will note that the above action causes data to be shifted serially into the register from the left-hand end, and shifted serially out at the right-hand end.

Shift registers of any length can be constructed simply by cascading *J-K*

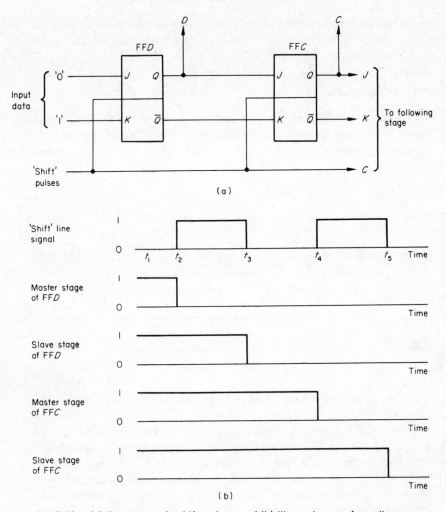

Fig. 7.13 (a) Two stages of a shift register, and (b) illustrative waveform diagrams

flip-flops in the manner shown in Fig. 7.13(a). Shift registers are used in a wide variety of applications ranging from temporary data storage to complete data transmission systems; in the latter type of system, the register is split between a master control station and several smaller outstations, possibly distributed over many miles in a large installation.

One industrial application of shift registers is the automatic rejection of faulty components in an automated production line. The 1's stored in the register represent the position of components on the production line that are satisfactory. If a test sequence indicates a fault on a component, a '0' is set into the appropriate flip-flop. This moves along the register as the component moves along the production line. The '0' in the register is used to inhibit further operations on the component, saving production time. At a suitable point the faulty component is rejected automatically by a circuit which recognizes the '0' in the register.

7.6 Ring counters

Ring counters are shift registers with the input to the register derived from the output flip-flop, as shown in Fig. 7.14. Assume initially that all the stages, except the first, are set to zero. The '1' in the first stage is then shifted one step

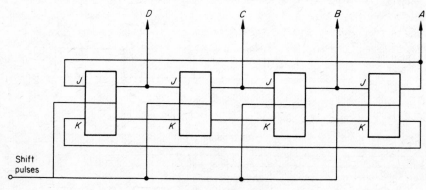

Fig. 7.14   A ring counter

along the register by each shift pulse. The pattern held in the counter is shown in Table 7.8.

Readers will note that, with the code sequence in Table 7.8, only four code groups are stored by the four bistables. The *maximum length* of the code patterns generated by this type of counter is N, where N is the number of flip-flops in the ring counter. Ring counters are uneconomic in terms of the electronic hardware required to produce a code sequence of a given length when compared with a pure binary counter since, in the latter case four flip-flops produce a code sequence of $2^4 = 16$ combinations. An advantage, however, of

Table 7.8

Pattern generated by a simple ring counter

	State of flip-flops			
Shift pulse	A	B	C	D
Initial condition	0	0	0	1
1	0	0	1	0
2	0	1	0	0
3	1	0	0	0
4	0	0	0	1 Repeat

the ring counter is that the process of decoding the stored data into decimal (and into many other codes) is very simple. The ring counter in Fig. 7.14 produces a variety of code sequences including three sequences of length four combinations (that in Table 7.8 being one of these), one sequence of length two code combinations, and two sequences of length one code combination (the all-0's and the all-1's combinations).

At the instant of switch-on, it is very probable that the ring counter will enter an incorrect code sequence. Additional logic circuitry is required to ensure that the correct code sequence is generated. The circuit in Fig. 7.15 is one such circuit which ensures that, whatever sequence the counter enters initially, the four-bit ring counter returns to the sequence in Table 7.8. It does so by feeding back to the J-input of FFC the NOR function of outputs A, B and C, which ensures that a '1' is presented to this input only when $A = B = C = 0$.

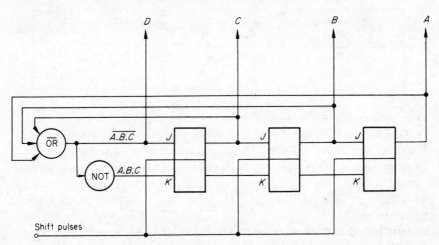

Fig. 7.15 One method of ensuring that the ring counter enters the code sequence in Table 7.8

7.7 Feedback shift registers

By feeding back more or less complicated functions of the state of the register to its input, shift registers can be made to generate a sequence of binary combinations. The sequence need not necessarily follow the normal pure binary counting sequence.

By feeding back the complement of the output it is possible, with a register with N stages, to generate pattern lengths of $2N$. The general arrangement is shown in Fig. 7.16. These are known as *twisted-ring counters* or *feedback shift registers* (FSR's). FSR's generate one walking or creeping code of length $2N$, in addition to other codes. Consider a FSR with complementary feedback similar to that shown in Fig. 7.16, having three stages A, B, and C. If the initial conditions are A_1, B_1, C_1 then the cycle in Table 7.9 is generated.

LOGIC CIRCUITS

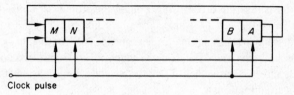

Clock pulse

Fig. 7.16   A feedback shift register using the complement of its own output as its input

Table 7.9

Main code of the three-stage FSR

Shift pulse number	State of the flip-flops		
	A	B	C
Initial condition	A_1	B_1	C_1
1	B_1	C_1	$\bar{A}_1$
2	C_1	$\bar{A}_1$	$\bar{B}_1$
3	$\bar{A}_1$	$\bar{B}_1$	$\bar{C}_1$
4	$\bar{B}_1$	$\bar{C}_1$	A_1
5	$\bar{C}_1$	A_1	B_1
6	A_1	B_1	C_1

Repeat

Thus if $A_1 = 0, B_1 = 0, C_1 = 1$ the pattern generated is a walking code of length $2N = 6$, shown in Table 7.10. Since there are $2^3 = 8$ possible combinations, another code of cycle length $2^N - 2N = 2$ exists. This is the code

$$A, B, C = 0,1,0 \rightarrow 1,0,1 \rightarrow 0,1,0$$

Strictly speaking the codes have no beginning and no end, and the code generated is dependent on the (arbitrary) state of the flip-flops at the commencement of the count. The code length of $2N$ is called the *main code* while the shorter code is called the *auxiliary code*. In general there may be an

Table 7.10

Walking code generated by a three-stage FSR

State of flip flops			BCD equivalent
A	B	C	
0	0	1	1
0	1	1	3
1	1	1	7
1	1	0	6
1	0	0	4
0	0	0	0
0	0	1	1

auxiliary code in addition to one or more main codes. The codes generated for a few values of N are given in Table 7.11.

In general there are 2^N possible combinations, and the length of each main code is 2^N. The number of main codes is the integral part of $2^N/2N$, and the remaining number is the length of the auxiliary code, e.g., if $N = 5$, $2^N = 32$, and $2N = 10$. There are therefore three main codes of length 10, and an auxiliary code of length two.

The main drawbacks to this form of counter are the short length of the main codes, and the fact that the counter may commence operation in any one of the codes. To ensure that the counter operates in only one code, the flip-flops have to be set initially to a pre-determined pattern.

Table 7.11

Number and length of the main auxiliary codes generated by FSR's

N	Number of main codes	Length of main codes	Length of auxiliary code
2	1	4	—
3	1	6	2
4	2	8	—
5	3	10	2
6	5	12	4

7.7.1   Linear FSR's

The length of the main code generated can be increased in some cases by feeding back a modulo-2, or NOT-EQUIVALENT function of the states of the register, as shown in Fig. 7.17.

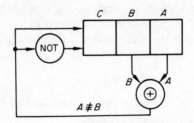

Fig. 7.17   A linear FSR using modulo-2 feedback

If the initial state is $A, B, C = 0,0,1$, the cycle in Table 7.12 ensues, bearing in mind that flip-flop C is set to the '1' state following the condition when $A \neq B$. This results in a code cycle of length 7, or $2^N - 1$. The missing state is 000 which, if generated, would terminate the cycle since C would not be set to '1' at the next shift pulse. In general, with modulo-2 feedback the cycle length is $2^N - 1$, and the condition in which all the outputs are zero should not be allowed to occur.

Table 7.12

Code cycle produced by Fig. 7.17

A	B	C
0	0	1
0	1	0
1	0	1
0	1	1
1	1	1
1	1	0
1	0	0
0	0	1

With the use of additional logic, the 'all 0's' state can be permitted to occur by ensuring that a '1' is injected into the register on the next shift pulse.

An alternative arrangement to that shown in Fig. 7.17, is obtained by feeding the lower input directly from the modulo-2 gate, and the upper input from the NOT gate. This results in a cycle length of $2^N - 1$, the 'all 0's' state being included in the cycle. The state $A, B, C = 1,1,1$, in this case, should not be allowed to occur.

The length and number of the cycles generated is dependent on the points from which the feedback is generated. For example, a five-stage FSR, with feedback taken from the 3rd and 5th or 2nd and 5th stages, has a main cycle length of 31 and a minor cycle length of one (the cycle 00000). If the feedback is taken from the 4th and 5th stages, it has a main cycle length of 21, and three minor cycles of lengths 7, 3, and 1 respectively. A general treatment of linear FSR's has been given by Elspas.[6]

A list of maximum sequence lengths for various lengths of shift registers and for different modulo-2 feedback connections is given in Table 7.13.

The name linear FSR is given to shift registers with modulo-2 feedback, since they are amenable to mathematical theory in much the same way as linear networks. If the state of an N-stage FSR at some point in the count is

$$\phi_n = N,M,L, \ldots, C,B,A$$

where N is the least significant term, and the modulo-2 feedback term is F, then the state of the counter after one shift pulse is

$$\phi_{n+1} = F,N,M, \ldots, D,C,B.$$

The function fed back can be any modulo-2 term, and is expressed mathematically as

$$F = nN \oplus mM \oplus \ldots \oplus bB \oplus aA$$

where $n, m, \ldots, b, a$ are coefficients which have a value of either '0' or '1'. If a four-stage linear FSR is used and feedback is obtained from the 3rd and 4th

Table 7.13

Feedback connections and sequence lengths

Number of stages	Sequence length	Feedback connections
2	3	1, 2
3	7	2, 3
4	15	3, 4
5	31	3, 5
6	63	5, 6
7	127	6, 7
8	255	4, 5 or 6, 8
9	511	5, 9
10	1023	7, 10
11	2047	9, 11
12	4095	6, 8 or 11, 12
13	8191	9, 10 or 12, 13
14	16383	4, 8 or 13, 14
15	32767	14, 15

stages, then $F = 0011$. ϕ_{n+1} can be related to ϕ_n by a transition matrix which contains the information given above, viz.:

$$
\begin{bmatrix} F \\ N \\ M \\ \vdots \\ C \\ B \end{bmatrix} =
\begin{bmatrix}
n & m & l & \cdots & c & b & a \\
1 & 0 & 0 & & 0 & 0 & 0 \\
0 & 1 & 0 & & 0 & 0 & 0 \\
\vdots & & & & & & \vdots \\
0 & 0 & 0 & & 1 & 0 & 0 \\
0 & 0 & 0 & & 0 & 1 & 0
\end{bmatrix}
\begin{bmatrix} N \\ M \\ L \\ \vdots \\ B \\ A \end{bmatrix}
$$

or

$$[\phi_{n+1}] = [T][\phi_n]$$

where $[T]$ is the transition matrix.

As a simple example, consider the three-stage linear FSR in Fig. 7.17 where $F = 011$. If the state of the register is $C, B, A = 1,1,0$ after 'n' pulses, then the two following states in the sequence are as calculated below.

$$
\begin{bmatrix} F \\ C \\ B \end{bmatrix}_{n+1} =
\begin{bmatrix} 0 & 1 & 1 \\ 1 & 0 & 0 \\ 0 & 1 & 0 \end{bmatrix}
\begin{bmatrix} 1 \\ 1 \\ 0 \end{bmatrix} =
\begin{bmatrix} 1 \\ 1 \\ 1 \end{bmatrix} =
\begin{bmatrix} C \\ B \\ A \end{bmatrix}_{n+1}
$$

$$
\begin{bmatrix} F \\ C \\ B \end{bmatrix}_{n+2} =
\begin{bmatrix} 0 & 1 & 1 \\ 1 & 0 & 0 \\ 0 & 1 & 0 \end{bmatrix}
\begin{bmatrix} 1 \\ 1 \\ 1 \end{bmatrix} =
\begin{bmatrix} 0 \\ 1 \\ 1 \end{bmatrix} =
\begin{bmatrix} C \\ B \\ A \end{bmatrix}_{n+2}
$$

This gives the states

A	B	C	Pulse number
0	1	1	n
1	1	1	$n + 1$
1	1	0	$n + 2$

This procedure allows any cycle to be fully calculated.

Linear FSR's can be used to generate binary sequences, known as *pseudo-random binary sequences* or PRBS. For example, the output sequence (in time order) from stage C of the FSR in Fig. 7.17 is 1, 0, 1, 1, 1, 0, 0. Using longer registers long PRBS's are generated. In the mid-1950's, PRBS were a mathematical curiosity, but are now important tools which have found a wide range of applications. They are used to encode and decode digital messages for the purpose of error-free transmission, and also to test control systems and to generate random number sequences.

7.7.2   Non-linear FSR's

Feedback shift registers which are not amenable to the above treatment are described as being non-linear. The main difference between linear and non-linear FSR's is that, in non-linear FSR's, the feedback terms are not necessarily modulo-2 functions, and multiple feedback loops may be used. Details of a wide range of binary sequence patterns is available in the literature.[7]

7.8   Parallel code conversion

Whilst computers and other logical networks operate with one or another form of binary code, man prefers to work in the decimal code. It is therefore necessary to be able to convert between binary codes and decimal. Conversion between binary codes is also necessary. For instance a unit-distance code may be employed by a *transducer* to measure the position of an object. If the information is to be processed by logic gates, it may be advisable to convert the information into some other code, say pure binary.

Many methods of code detection are employed in practice. To illustrate this consider the detection of the sequence $A, B, C, D = 1,0,1,0$. If the number is to be defined uniquely then it is necessary to define each term as shown in Fig. 7.18(a). The output from the AND gate is zero unless $A, \bar{B}, C$ and $\bar{D}$ are all 1's, when it is '1'. The NAND gate gives an output of '0' when the 1010 state is detected, while the NOR gate gives a '1' with input lines $\bar{A}, B, \bar{C}, D$. In certain cases it may not be necessary to define the whole number uniquely. For example if the 1's only occur in the configuration given under one condition, then it is only necessary to detect the 1's. In the case of the AND gate in Fig. 7.18(a), input lines A and C only are required.

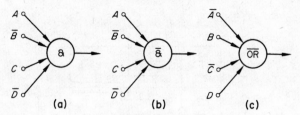

Fig. 7.18 Methods of detecting a code combination

Examples of convertor design are considered below, and it will be seen that sequences with redundant states allow some simplification to be made.

Example 7.1: 8421 BCD to decimal convertor.

Solution: The code is shown in Table 7.14 and the Karnaugh map in Fig. 7.19, the 'can't happen' conditions being marked with X's. Each cell is indicated by its decimal value, and it is defined by grouping it with the largest number of X's possible. Cells zero and unity are unique and must be defined in terms of the

Table 7.14

Code for 8421 BCD convertor

Decimal number	A (8)	B (4)	C (2)	D (1)
0	0	0	0	0
1	0	0	0	1
2	0	0	1	0
3	0	0	1	1
4	0	1	0	0
5	0	1	0	1
6	0	1	1	0
7	0	1	1	1
8	1	0	0	0
9	1	0	0	1
10–15	X	X	X	X

X = 'can't happen' condition

four binary variables A, B, C, D. The other cells can be grouped with one or more X's. The logic to give the decimal outputs is given in Table 7.15.

The block diagram of the decoding network, using AND gates, is shown in Fig. 7.20. The state of the count is given by a '1' at the output of the appropriate gate. Had NAND gates been used, the state of the count would be indicated by the gate which had a '0' at its output. The device matrix (see section 4.9) shows that the positive logic AND function is carried out if NOR gates are used with positive output and negative input logic. That is, the complements of the AND inputs are required if NOR gates are to be used.

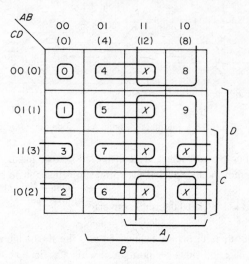

Fig. 7.19   Karnaugh map for example 7.1

This technique can be applied to an 8421 BCD counter so that an output pulse is obtained after a pre-determined number of input pulses have been received, the required number of input pulses being set by a switch. An electrical circuit fulfilling this requirement is shown in Fig. 7.21, illustrating the condition for an output on the count of 8_{10}. By interchanging the input lines to each switch, a NOR gate can be used at the output. One application of this circuit is to a batch counter, when an output exists after a pre-determined number of items have passed one point. This pulse can be used to initiate some other process, such as packaging, and also to reset the counter to zero to commence the cycle again.

Example 7.2: Decimal to a walking code convertor.

Table 7.15

Logic required for 8421 BCD to decimal convertor

Decimal number	Logic
0	$\bar{A} . \bar{B} . \bar{C} . \bar{D}$
1	$\bar{A} . \bar{B} . \bar{C} . D$
2	$\bar{B} . C . \bar{D}$
3	$\bar{B} . C . D$
4	$B . \bar{C} . \bar{D}$
5	$B . \bar{C} . D$
6	$B . C . \bar{D}$
7	$B . C . D$
8	$A . \bar{D}$
9	$A . D$

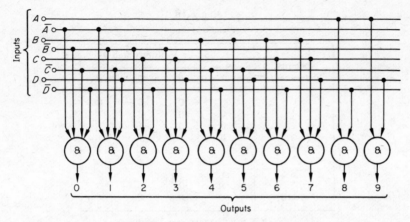

Fig. 7.20   Logic network for example 7.1

Solution: The code cycle of the walking code is given in Table 7.16. Ten input lines are available, and these are to be converted into four outputs, A', B', C', and D'. The decoding logic is obtained by inspecting the relevant rows of Table 7.16. An output exists at the A'-line when either the 4_{10} OR 5_{10} OR 6_{10} OR

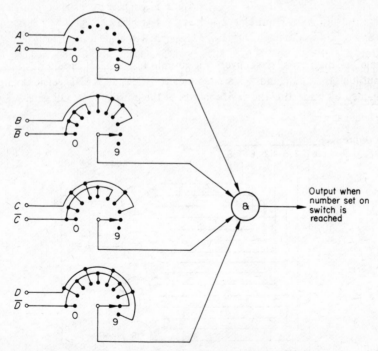

Fig. 7.21   A simple circuit to give an output after a given number of pulses have been recieved

Table 7.16

A 'walking' code

Decimal number	States of output lines			
	A'	B'	C'	D'
0	0	0	0	0
1	0	0	0	1
2	0	0	1	1
3	0	1	1	1
4	1	1	1	1
5	1	1	1	0
6	1	1	0	0
7	1	0	0	0
8–15	X	X	X	X

X = 'can't happen' condition

7_{10} lines are energized. This is expressed logically as

$$A' = 4_{10} + 5_{10} + 6_{10} + 7_{10}.$$

The following logical relationships hold for the remaining outputs:

$$B' = 3_{10} + 4_{10} + 5_{10} + 6_{10}$$
$$C' = 2_{10} + 3_{10} + 4_{10} + 5_{10}$$
$$D' = 1_{10} + 2_{10} + 3_{10} + 4_{10}.$$

The block diagram of the convertor is given in Fig. 7.22. It should be noted that input lines 0_{10}, 8_{10}, and 9_{10} are redundant. NOR and NAND versions can be obtained by inspecting the device matrix of the positive logic OR gate.

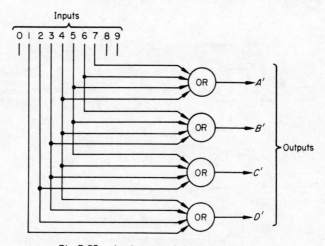

Fig. 7.22   Logic network for example 7.2

Example 7.3: A non-decimal to non-decimal code convertor.

Solution: The general principles have been outlined in examples 7.1 and 7.2; it remains merely to collect the methods together. As an example we will consider the design of a network to convert from the 8421 BCD code in Table 7.14, to the walking code in Table 7.16. The Karnaugh map for the BCD-to-decimal conversion is shown in Fig. 7.19, and the logical expressions for the walking code outputs, A', B', C', and D', are given in the solution to Example 7.2. To obtain the overall logical relationships for the convertor, separate Karnaugh maps are drawn for each output line, in Fig. 7.23. It is assumed that input conditions corresponding to 8_{10} and 9_{10} cannot occur ('can't happen' conditions).

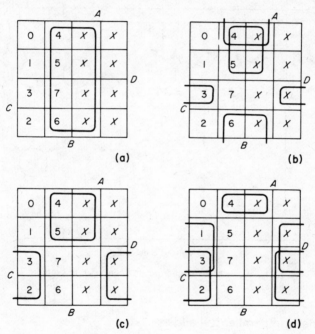

Fig. 7.23 (a), (b), (c), and (d)   Karnaugh maps for the convertor in example 7.3

The logical expression for output A', obtained by grouping the appropriate decimal terms in Fig. 7.23(a), is

$$A' = B.$$

The logical expressions for outputs B', C', and D', from Fig. 7.23(b), (c), and (d), respectively, are

$$B' = B \cdot \bar{C} + B \cdot \bar{D} + \bar{B} \cdot C \cdot D$$
$$C' = B \cdot \bar{C} + \bar{B} \cdot C$$
$$D' = \bar{B} \cdot C + \bar{B} \cdot D + B \cdot \bar{C} \cdot \bar{D}.$$

The logical block diagram of the convertor is shown in Fig. 7.24. The number of gates used in the convertor is three less than is required in a code convertor in which the intermediate decimal values are available.

Each gate in Fig. 7.24 can be replaced by a NAND gate, since the output logic is in the form of the logical sum of logical products. A NOR network can be used if the input and output logical levels are inverted.

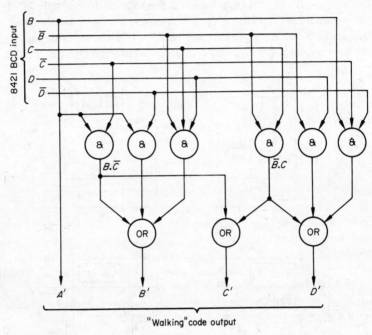

Fig. 7.24   Logic block diagram of the convertor in example 7.3

7.9   Serial and parallel full-adders

In pocket calculators and in computers, operands which are to be added together are stored in registers, and the addition process is carried out automatically.

The basis of the serial full-adder is shown in Fig. 7.25. The numbers, M and N, to be added together are stored in registers M and N, respectively. Before operations commence, the value stored both by the sum register, S, and the carry store are set to zero. At the commencement of operations, the least significant bits of operands M and N are presented to the full-adder and, since this circuit contains only combinational logic, the sum and carry bits of these inputs appear simultaneously at outputs S and C_O. The first clock (shift) pulse causes the signal at output S to be shifted into the most significant location of register S. It also causes the state of output C_O to be transferred into the D-type

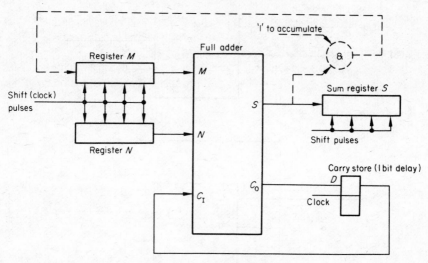

Fig. 7.25　A serial full adder

carry store; at the end of the clock pulse cycle, the carry bit associated with the addition of the least significant bits of M and N appears at the output of the carry store, and is applied to the carry-in input, C_I, of the full-adder. Thus, the carry store introduces the required one-bit delay for the full addition process. The application of the correct number of shift pulses results in the sum of the two operands being stored in register S; the number in this store is progressively shifted to the right as clock pulses are applied, so that the least significant bit of the sum is finally stored in the extreme right-hand location of register S.

In practice, register M is used for the dual purpose of storing the value of operand M, and also for storing the final sum. This is brought about by means of the 'accumulate' loop shown in broken line in Fig. 7.25. If a logic '1' is applied to the AND gate in this loop then, as the original data in register M is shifted right, so the sum of the two numbers is shifted into the opposite end of the register. When the original number stored in register M has been completely shifted out, this register contains the sum of M and N. The sum register is now surplus to requirements.

In a parallel full-adder, the binary digits are applied simultaneously to the circuit. The two stages of a parallel full-adder are shown in Fig. 7.26; in addition to a full-adder for each pair of bits, three registers are required to store the two operands and the sum (two are needed if one of the registers is used in the 'accumulate' mode, as in Fig. 7.25). It is evident that parallel full adders are more complex than are serial adders, but they have the advantage of speed since the full addition is carried out simultaneously.

Binary subtraction, both in the serial and the parallel modes, is carried out using complement addition techniques. The circuits used for the addition part of this process are as shown in Figs. 7.25 and 7.26.

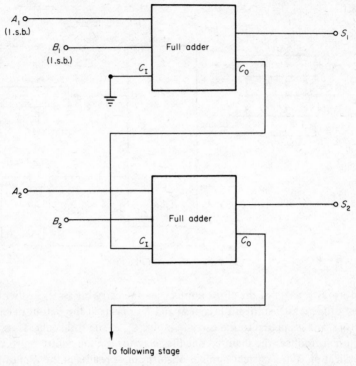

To following stage

Fig. 7.26   Two stages of a parallel full adder

Problems

7.1. What is meant by the term *synchronous* as applied to counters? Discuss the need for synchronizing pulses or clock pulses in a synchronous counting system.

7.2. Write down a BCD code with a weighting of 3321, and design a synchronous counter employing *J-K* flip-flops which generates this code.

7.3. Design a counter which generates the Gray code.

7.4. Design a counter using (a) *J-K* flip-flops, (b) *T* flip-flops, and (c) *S-R* flip-flops, which counts in the following sequence.

$$
\begin{array}{c}
0000 \\
0001 \\
0010 \\
0100 \\
1000 \\
1001 \\
0110 \\
1111 \\
\hline
0000 \\
\text{etc.}
\end{array}
$$

7.5. What is meant by the expression *shifting register*? Illustrate, with waveform diagrams, the operation of a shifting register.

7.6. Write down all the possible code patterns for a five-stage feedback shift register.

7.7. A linear feedback shift register has four stages, designated A, B, C, and D. The feedback signal to stage D is generated from stages A and C. Compute the code sequence generated.

7.8. A position encoder generates the following unit-distance decimal code, known as the Petherick code. Design a convertor, using only NOR gates, to convert the output into the pure binary code.

```
0101
0001
0011
0010
0110
1110
1010
1011
1001
1101
────
0101
etc.
```

References

1. MARCUS, M. P., 'Cascaded binary counters with feedback', *Trans. Inst. elect. Electron. Engrs.*, **EC-12**, 4, 361, 1963.
2. CALDWELL, S. H., *Switching circuits and logical design*, John Wiley.
3. PHISTER, M., *Logical design of digital computers*, John Wiley.
4. BIWAS, N. N., 'The logic and input equations of flip-flops', *Electron. Engng.*, **38**, 2, 107, 1966.
5. DEAN, K. J., 'The design of parallel counters using the map method', *J. Inst. Elect. Radio. Engrs.*, **32**, 159–62, 1966.
6. ELSPAS, B., 'The theory of autonomous linear sequential networks', *Trans. Inst. Radio Engrs.*, **CT-6**, 45, 1959.
7. HOFFMANN de VISME, G., *Binary Sequences*, E.U.P.

8. Electronic switching devices and circuits

8.1 Semiconductor terminology

A *semiconductor* is a material whose conductivity, at room temperature, is between that typical of conductors and insulators.[1] Germanium and silicon are the most commonly used semiconductor materials. Pure semiconductors become perfect insulators at absolute zero temperature, and their conductivity rises with room temperature. This is known as *intrinsic conductivity* and is regarded as an imperfection in the material. By doping the pure semiconductor with a controlled amount of impurity (a few parts per million), another form of conductivity, known as *extrinsic conductivity*, is introduced.

The impurities introduced are referred to either as *p-type* or *n-type*. Semiconductors with a p-type impurity contain mobile positive charge carriers (known as *holes*), while the n-type impurity semiconductor contains mobile negative charge carriers (*electrons*). The electrical charge associated with the 'hole' and the electron are equal and opposite, and if a hole and an electron combine, the net electrical charge is zero.

Flow of current in a n-type material is largely due to the movement of electrons through it, and they are described as *majority charge carriers*. If positive charge carriers (holes) appear in an n-type semiconductor they also constitute flow of current, and are described as *minority charge carriers*. In a p-type material holes are the majority charge carriers, and electrons are the minority charge carriers.

8.2 P-N junction diode

A p-n junction diode comprises a semiconductor crystal with both p- and n-type regions, the two being joined by molecular bonds, shown in Fig. 8.1(a), together with its circuit symbol (b). It is found that current flow occurs only when the

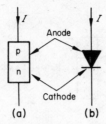

Fig. 8.1   (a) The basic physical arrangement of the p-n junction diode, and (b) its circuit symbol

p-type anode is positive with respect to the n-type cathode. In this state the diode is said to be *forward biased* and *forward conduction* takes place, the corresponding part of the characteristic being shown in Fig. 8.2.

When the anode is negative with respect to the cathode, the diode is said to be *reverse biased*, and only a minute leakage currents flows. This is known as the *reverse blocking* state, when flow of current is blocked. A significant increase in reverse bias results in electrical breakdown of the diode, when it reverts to its second conducting state, known as *reverse conduction*. In semiconductor devices used as conventional diodes this usually results in a catastrophic failure, but diodes known as *Zener*, or *avalanche diodes* are operated on the reverse breakdown region.

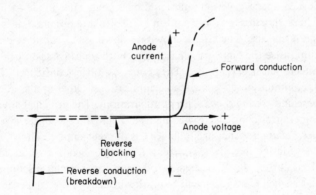

Fig. 8.2   Static characteristic of the p-n junction diode

8.3   Junction transistors

An n-p-n junction transistor is formed in a single semiconductor crystal, and has two n-regions and a p-region, as shown in Fig. 8.3(a). The n-p-n transistor is said to be a *bipolar* device since conduction is through two types of semiconductor material, and is carried by two types of charge carrier. The circuit symbol is given in Fig. 8.3(b), the arrow on the emitter giving the conventional direction

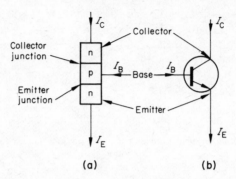

Fig. 8.3   The physical representation of the n-p-n transistor, and (b) its circuit symbol

of flow of current through the device, electron flow being in the opposite direction. The n-type *emitter* is the source of current carriers (electrons), which are divided between the *base* and *collector* regions. The name collector implies that it is the region at which most of the current carriers leaving the emitter are collected. The name base region dates back to the construction of early devices, and is regarded as the control electrode for the purposes of this book.

The p-n junction between the base and emitter regions is known as the *emitter junction*, and that between the base and collector regions as the *collector junction*. At first glance it appears that the simple equivalent electrical circuit of the transistor comprises two p-n junction diodes connected anode-to-anode. When the characteristics for the circuit in Fig. 8.4(a) are obtained, it is found that this is not the case. The circuit shown is known as the *common-emitter* configuration, since the emitter is common to both input (base-emitter) and output (collector-emitter) supplies. The common-emitter configuration is the mode most commonly used in switching applications. Other modes are the *common-base* and *common-collector* configurations. The principal advantage of the common-emitter configuration is the resulting high power gain.

The *output characteristic*, in Fig. 8.4(b), is of great value to circuit designers. It shows the variation in collector current, I_C, with collector-emitter voltage, V_{CE}, and base current I_B. With zero base current (base circuit disconnected), it is found that a small leakage current I_{CE0} flows between the collector and emitter. This leakage current lies between about 50 nA and a few microamperes. In this state the transistor is said to be 'OFF', in that it is approximately equivalent to a switch that is 'off' or 'open'.

By increasing I_B, at a constant value of V_{CE}, it is found that the collector current increases roughly in proportion to the base current, according to the relationship (neglecting leakage current)

$$I_C = h_{FE}I_B.$$

A circuit diagram, known as the *equivalent circuit*, which embodies this fact is shown in Fig. 8.4(c). The constant current generator shunting the collector diode generates the current $h_{FE}I_B$, and resistor R in the base line represents the ohmic resistance of the base region itself. Figure 8.4(c) is an equivalent circuit that is suitable for low frequencies, and is used here only to explain the circuit operation under steady-state conditions.

Parameter h_{FE} is the static value of the forward (signified by suffix F) current transfer ratio, I_C/I_B, in the common-emitter (signified by suffix E)

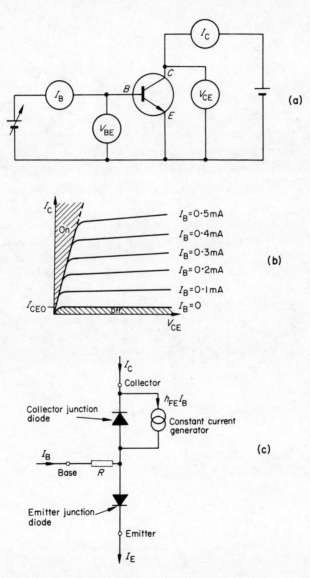

Fig. 8.4   (a) A circuit used to determine the output characteristic, (b), of the n-p-n junction transistor. (c) The simplified equivalent circuit

configuration. The letter h refers to a set of hybrid parameters which are used to define the operation of the transistor. This parameter is dependent not only on the collector current at which it is measured, but also on the value of V_{CE}, and the semiconductor junction temperature T_j. It is therefore defined at a given value of V_{CE} and I_C, usually at 25°C. Typical curves giving variations in h_{FE} with I_C and T_j at constant V_{CE} are shown in Fig. 8.5.

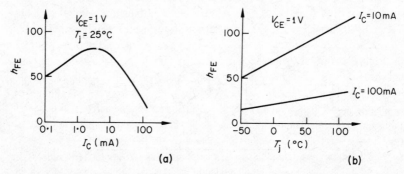

(a)   (b)

Fig. 8.5   Variation of hFE with (a) I_C, and (b) junction temperature

By injecting a large base current, the transistor *saturates*, and is said to be *bottomed* or *turned-on*. The region of the output characteristic used in this state is labelled ON in Fig. 8.4(b). The problem here is what is the value of h_{FE} to be used in calculations? This problem is illustrated in more detail in Fig. 8.6. When V_{CE} is 1 V, the value of h_{FE}, for a base current of 1 mA, is h_{FE} = 30 mA/1 mA = 30. For the same value of base current, for values of V_{CE} of 0.5 V and 0.25 V, h_{FE} is 25 and 12.5 respectively. For a V_{CE} of 1 V, the transistor is not fully turned ON, and the value of 30 for h_{FE} is clearly not applicable for a base drive of 1 mA. When I_C = 25 mA the transistor is

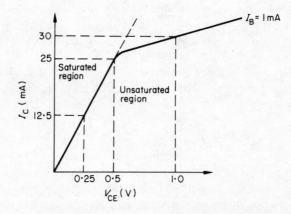

Fig. 8.6   Static characteristic in the region of saturation

approaching the ON state, but due to the spread of parameters other transistors of the same type may not saturate with I_B = 1 mA. To ensure that all transistors of the same type saturate, a working point of V_{CE} = 0·25 V, I_C = 12·5 mA, with a base current of 1 mA, should be aimed at, i.e., an h_{FE} of 12·5.

Manufacturers quote a number of values of h_{FE}, at points on the characteristics, at various values of collector current with V_{CE} constant, usually at 1 V. The true figure can only be obtained from the characteristics of the transistor to be used, but a suitable value may be obtained more rapidly from the data sheets as follows. The manufacturer specifies the maximum voltage existing across the transistor, $V_{CE(sat)}$, when it is in the saturated state with given collector and base currents. The value of h_{FE} to be used can be deduced from these figures. For example, if a silicon switching transistor has a specified $V_{CE(sat)}$ of 125 mV with I_C = 10 mA and I_B = 1 mA, the resulting figures of $h_{FE(sat)}$ = 10 is a conservative figure for that type of transistor, and may be used in calculations. The parameter $h_{FE(sat)}$ is defined here as the static forward current ratio when the transistor is fully saturated. In general, the value of $h_{FE(sat)}$ is lower than the unsaturated value of h_{FE}, and is determined by the base current and the collector current (which is limited by external factors such as the value of the load resistance).

An alternative type of bipolar transistor, the p-n-p transistor, is also available. This comprises an n-region (the base) between two p-regions (the emitter and collector). With this type of device, negative collector and base potentials are used, otherwise the principle of operation is broadly similar to the n-p-n type.

The features of both types of transistor which make them suitable as switching devices are:

1. They are physically small, and the power dissipation in both the ON and OFF states is low.
2. Their cost is low, and the advent of integrated circuits has accelerated the reduction in cost per logical function.
3. The collector current in the saturated state is controlled by a much smaller base current (I_B = I_C/h_{FE}).

Their main disadvantages are:

1. The device is never truly OFF, since a small leakage current flows.
2. It can never be fully turned ON, since the saturation voltage $V_{CE(sat)}$ is finite.

8.4   Resistor-transistor NOT gate

The circuit in Fig. 8.7(a) fulfils the requirements of a NOT gate. When the base is connected to the zero potential line (logical '0'), the base current is zero and the transistor is in the OFF state. The collector current is practically zero, and the output potential is high (logical '1'). When the base potential is high

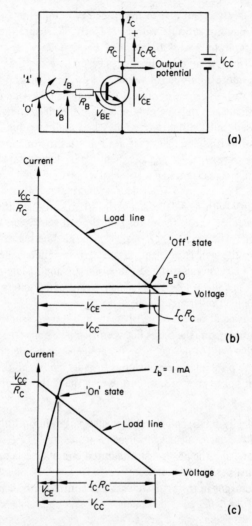

Fig. 8.7   (a) Simple NOT gate using a transistor circuit; (b) and (c) show the OFF and ON operating states, respectively

(logical '1'), the transistor is turned ON, and V_{CE} is very small (logical '0'), typically 0·2–0·5 V.

The collector current in Fig. 8.7(a) can be evaluated by determining the conditions under which the following are satisfied simultaneously:

(a)   Current through the resistor = current through the transistor

(b) $$V_{CC} = V_{CE} + I_C R_C.$$

A precise mathematical solution can be obtained, but an easier and quicker method results from a simple graphical solution as follows. The collector potential is given by the equation

$$V_{CE} = V_{CC} - I_C R_C$$

or

$$I_C = \frac{V_{CC}}{R_C} - \frac{V_{CE}}{R_C}.$$

This is the equation of a straight line of slope $-1/R_C$, terminating at points $V_{CE} = V_{CC}$ when $I_C = 0$, and $I_C = V_{CC}/R_C$ when $V_{CE} = 0$. The line is known as the *load line* and is shown in Figs. 8.7(b) and (c). The intersection of the load line with the appropriate output characteristic gives the point where the transistor and resistor currents are equal. The voltage across the transistor, V_{CE}, and across the resistor, $I_C R_C$, are then read off the output characteristic as shown.

When the switch in the base line connects the base to '0', $V_{CE} \simeq V_{CC}$ and $I_C \simeq 0$. When the input switch connects the base to the logical '1' level, when $V_B = V_{CC}$, a finite base current flows, given by

$$I_B = \frac{V_{CC} - V_{BE(sat)}}{R_B}$$

where $V_{BE(sat)}$ is the base-emitter voltage when the transistor is in the saturated state. $V_{BE(sat)}$ is small, compared with V_{CC}, and it may be neglected, giving

$$I_B \simeq V_{CC}/R_B.$$

By applying adequate base current (referred to as base 'drive'), the transistor becomes saturated, and the collector potential falls to $V_{CE(sat)}$.

A simple design procedure for a transistor NOT gate is illustrated as follows. Consider transistor Q1, in Fig. 8.8(a), which has M similar circuits connected to its collector. The maximum value of M that Q1 can supply is known as the *fan-out* of the circuit. When the collector voltage of Q1 is 'high', the circuit provides base drive to transistors Q2, Q3, etc; consequently the drive circuit acts as a current source, and is described as a *current sourcing* logic gate. In general, bipolar logic circuits of the OR and NOR types are current sourcing gates. The worst operating state occurs when the input to Q1 is '0', and the maximum fan-out, M_{max}, is connected. In this case, the collector potential of Q1 is 'high', and it must be possible to draw sufficient current through R_C to drive all the connected gates into saturation. If the leakage current of Q1 can be neglected, the current through R_C when Q1 is OFF is

$$I_C = \frac{V_{CC} - V_{BE(sat)}}{R_C + R_B/M}.$$

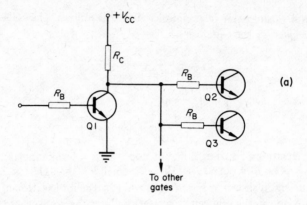

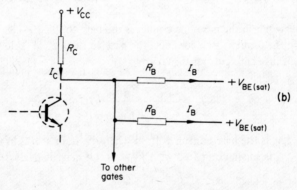

Fig. 8.8 (a) and (b)   Worse-case design of the NOT gate in Fig. 8.7, with a fan-out of M

This operating condition is illustrated in Fig. 8.8(b). Neglecting $V_{BE(sat)}$, which is usually small compared with V_{CC}, the equation becomes

$$I_C \simeq \frac{V_{CC}}{R_C + R_B/M}$$

The base current supplied to each of the connected transistors Q2, Q3, etc., is

$$I_B = I_C/M = \frac{V_{CC}}{MR_C + R_B}.$$

But $I_C = h_{FE(sat)}I_B$, therefore

$$I_B = \frac{I_C}{h_{FE(sat)}} = \frac{V_{CC}}{MR_C + R_B}.$$

If $V_{CE(sat)}$ is small, then $I_C \simeq V_{CC}/R_C$ when the transistor is ON.

Hence

$$\frac{V_{CC}}{R_C h_{FE(sat)}} = \frac{V_{CC}}{MR_C + R_B}$$

or

$$M = h_{FE(sat)} - R_B/R_C.$$

Clearly M has a maximum value, with a given value of $h_{FE(sat)}$, when R_B/R_C is minimum. There are limits to the values of R_B and R_C that may be used, since a large value of R_C limits the current that may be drawn from the supply, and a small value of R_B results in an excessive base current demand.

For a typical silicon switching transistor, $V_{CE(sat)} = 0\cdot25$ V at a collector current of 10 mA, and base current of 1 mA. Here $h_{FE(sat)}$ is 10, and if values of $R_B = R_C = 1$ kΩ, then $M_{max} = 10 - 1 = 9$, i.e., nine similar NOT gates may be connected to the collector of any transistor. The above calculations assumed that V_{CC} was constant, and the resistance values were not subject to variation in value. In practice, a tolerance must be allowed on these values, in addition to which h_{FE} will vary between transistors of the same type. Thus in the above case, if $h_{FE(sat)}$ lies between 9·5 and 12·5, and the resistors have a tolerance of 10%, the worst case occurs when $h_{FE(sat)} = 9\cdot5$, $R_B = 1\cdot1$ kΩ, and $R_C = 0\cdot9$ kΩ. This gives a new maximum theoretical fan-out of 8·28. Since an integral number of circuits only can be connected, the fan out is reduced to eight.

8.5   Saturated operation

A transistor with an $h_{FE(sat)}$ of 10 may, typically, have a working h_{FE} in the unsaturated region of 50. The approximate equivalent circuit of an n-p-n transistor in the unsaturated region, with $h_{FE} = 50$ at a collector current of 5 mA ($I_B = 0\cdot1$ mA) with a 1 kΩ collector load, is shown in Fig. 8.9(a). In the unsaturated region, the emitter diode is forward biased, and the constant current generator allows the collector current of 5 mA to flow in the load. The emitter current is $I_C + I_B = 5 + 0\cdot1 = 5\cdot1$ mA, resulting in a forward voltage drop of, say, 0·55 V across the emitter diode. The p.d. across the 1 kΩ load is 5 mA x 1 k$\Omega = 5$ V, and the collector potential is $10 - 5 = 5$ V. The collector diode has, therefore, a reverse bias of $5 - 0\cdot55 = 4\cdot45$ V across it. A further increase in base current results in an increase in collector current, and a change in h_{FE}, here a reduction since the transistor is approaching saturation. The increased base and collector currents result in an increased emitter current, and increased forward voltage drop in the emitter diode.

In Fig. 8.9(b), a base current of 0·47 mA results in an h_{FE} of 20, giving a

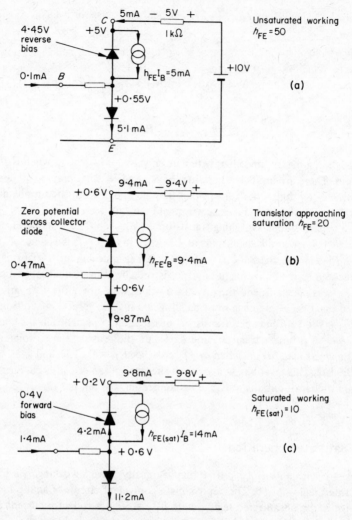

Fig. 8.9   Operation of a junction transistor (a) in the unsaturated region, (b) on the verge of saturation, and (c) in the saturated region

collector current of 9·4 mA and an emitter current of 9·87 mA. The forward voltage drop across the emitter diode increases to 0·6 V as a result of this. Since 9·4 mA flows in the collector circuit, the p.d. across the 1 kΩ resistor is 9·4 V. That is, the collector potential is $10 - 9·4 = 0·6$ V. The reverse bias across the collector diode has now fallen to zero, and the transistor is on the verge of saturation.

Any further increase in base current results in a further reduction in the static forward current transfer ratio, illustrated in Fig. 8.9(c). Here $I_B = 1·4$ mA and $h_{FE(sat)}$ is 10, resulting in a current of 14 mA through the constant current generator. Under these conditions the collector potential is of the order of

0·2 V, hence the current in the 1 kΩ resistor is $(10 - 0·2)/1 = 9·8$ mA. The difference between the external and internal collector currents $(14 - 9·8 = 4·2$ mA) circulates through the collector diode, which has now become forward biased. The effect of the forward bias on the collector diode is to reduce the collector potential below that of $V_{BE(sat)}$ (to 0·2 V in this case).

By using saturated transistors, one of the logical levels is practically equal to zero potential, while the other logical level can approach V_{CC}. The effective resistance between the collector and emitter in the saturated state is known as the *saturation resistance*, $r_{CE(sat)}$. In Fig. 8.9(c), $r_{CE(sat)} = 0·2$ V/9·8 mA = 20·4 Ω.

One problem of using the circuits in Figs. 8.8 and 8.9 is that $V_{CE(sat)}$ rises, and $V_{BE(sat)}$ falls, with temperature. This means that a rise in temperature will reduce the margin by which one transistor in the ON state can hold a number of connected transistors in the OFF state. Offset against this is the fact that h_{FE} rises with temperature.

The principal advantage of saturated switching circuits is their simplicity. The main disadvantage is that the switching speed is reduced, since, during the switch-off period, the base region must be swept clear of base charges, which takes a finite time.

8.6   Transistor turn-on

The effect of a small change in base current on collector current can be measured at any given frequency. A graph of the small-signal forward current transfer ratio, h_{fe}, over a range of frequencies, to a logarithmic base, is shown in Fig. 8.10. The value of h_{fe} is constant over a wide range of frequency, falling off

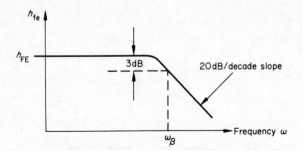

Fig. 8.10   Variation of h_{FE} with frequency

at the high-frequency end of the spectrum at the rate of approximately 20 dB/decade. At the low frequency end it assumes the value h_{FE}. It can be shown[2] that at some frequency, ω

$$h_{fe} = \frac{h_{FE}}{1 + j\omega/\omega_\beta}$$

where $j = 1 \underline{/90°}$, and ω_β is the 'corner' or 'cut-off' frequency, at which the gain
is 3 dB below the low-frequency gain.

The transistor, in the common-emitter mode, thus displays the frequency
response characteristic of a single time-lag network with a time-constant
$T = 1/\omega_\beta$. It follows that, for a step change in base current from zero to I_{B1}, the
collector current I_{C1} at any time t is given by

$$I_{C1} = h_{FE}I_{B1}(1 - e^{-t/T}). \tag{8.1}$$

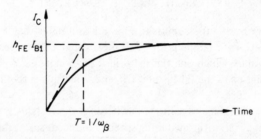

Fig. 8.11   Transient variation of I_C with a step change in base current

The general response of this equation is shown in graphical form in Fig. 8.11. An
increased base drive results in a higher prospective value of collector current, as
shown in Fig. 8.12. In practical circuits the collector current cannot exceed
V_{CC}/R_C. Any prospective current above this value (known as *overdrive* current)
must circulate through the internal collector diode of the transistor, and does
not appear in the collector circuit. Overdrive directly results in a more rapid
turn-on of the transistor, a reduction from t_1 to t_3 in Fig. 8.12, for an increase
in base current by a factor of three.

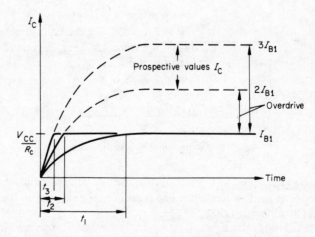

Fig. 8.12   Reduction of turn-on time with overdrive

The actual collector current waveform, for a step change in base current, is shown in Fig. 8.13. The *rise-time*, t_r, is defined as the time taken for the collector current to rise from 10 to 90% of its final value. During this time the current rises along an exponential curve, as described above. At the instant of turn-on, base current begins to flow, but it takes a finite time for the current carriers to spread across the base region to initiate the turn-on mechanism. This delay is described as the *turn-on delay* time t_d. The total time required to turn the transistor ON is known as the *turn-on time*, t_{on}, where

$$t_{on} = t_d + t_r$$

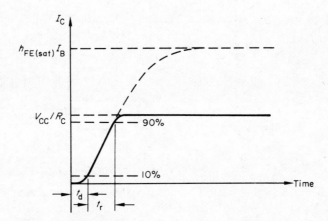

Fig. 8.13   Definition of turn-on delay t_d, and rise-time t_r

8.7   Transistor turn-off

When the transistor is saturated, both junctions are forward biased. The time taken for the overdrive current to be swept out of the collector junction, when turning the transistor OFF, is known as the *storage time* t_s, shown in Fig. 8.14. When these current carriers have been swept out, I_C falls in an exponential manner. The *fall-time* or *turn-off transition*, t_f, of the collector current, is taken as the time for the current to fall from 90 to 10% of the initial steady value. The total time required to turn the transistor OFF is known as the *turn-off time*, t_{off}, where

$$t_{off} = t_s + t_f$$

8.8   Propagation delay

A parameter used in specifying the switching performance of logic gates is the average time taken for the signal to propagate from the input to the output of the gate. This parameter is known as the *propagation delay*, t_{pd}, and is defined

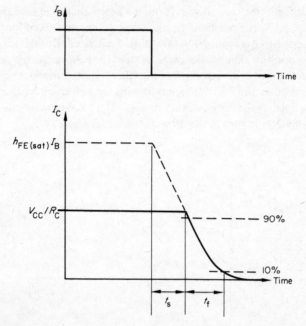

Fig. 8.14   Effect of overdrive on the turn-off time of the transistor

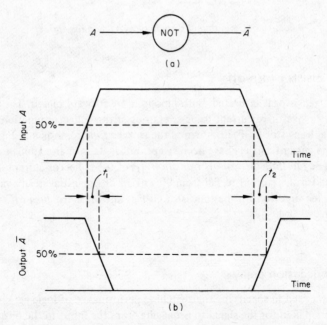

Fig. 8.15 (a) and (b)   Propagation delay of a gate

for the inverting gate in Fig. 8.15(a) in terms of the waveform diagrams in diagram (b). The propagation delay is specified at the 50 percent voltage levels of the input and output signals, and is given by

$$t_{pd} = (t_1 + t_2)/2$$

where t_1 and t_2 are defined in Fig. 8.15(b). The propagation delay depends on the type of logic circuit in use and on its loading, and may have a value less than one nanosecond in some electronic circuits. In fluid logic circuits, the value of t_{pd} may be several milliseconds.

8.9   Methods of reducing the switching time

It has been shown that a large base current is necessary to reduce the overall turn-on time, but it has the effect of prolonging the storage time. The simple circuit, shown in Fig. 8.16, improves the overall switching performance. When V_B is increased from zero, the capacitor draws a large charging current, ensuring that the base current is initially high. The transistor is rapidly forced into saturation by this artifice. When the capacitor becomes fully charged the base current falls to a value $I_R = (V_B - V_{BE(sat)})/R$. Providing that this value is sufficiently large to maintain the transistor in a state of saturation, the overdrive and hence storage time are reduced. The capacitor discharges rapidly when the

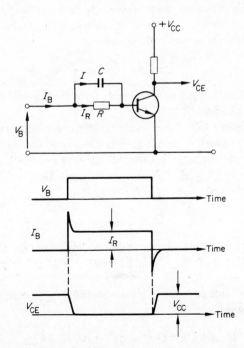

Fig. 8.16   Reduction of switching time by use of an *R-C* circuit

input voltage is reduced to zero, drawing a reverse current from the base of the transistor. This ensures that the charge carriers in the base region are rapidly swept out when the applied voltage is reduced to zero.

So far the transistor has been regarded as a current-controlled device. In many switching applications, particularly those where capacitors are used to improve response time or as trigger devices, it is more convenient to regard the transistor as a *charge-controlled* device.[3,4] Charge control theory is based on the concept that the charge built up in the base region by minority carriers is the primary control mechanism.

There are three primary charge storage mechanisms associated with the base region of the transistor. One is due to the transit of minority carriers, through the base region, from the emitter to the collector. The charge stored by this effect is

$$Q_B = I_C \tau_C$$

where τ_C is a time constant which depends on the base width, and is representative of the transit time of the minority carriers through the base region. For the treatment adopted here, it is convenient to define the operation when the transistor is on the edge of saturation, i.e., when

$$Q_B = I_C \tau_{C0}.$$

When working in the active region, charge Q_V is stored in the collector depletion layer capacitance, C_{TC}, when the voltage across it changes by δV_{CB}. This is the second charge storage mechanism

$$Q_V = K C_{TC} \, \delta V_{CB}$$

where K is a constant having a value between 1 and 2, depending on the construction of the transistor.

At the onset of saturation the collector junction becomes forward biased. When this occurs, any further charge carriers which are injected into the base cannot enter the collector, so increasing the charge stored in the base region. This is represented by Q_{BX}, the third charge storage effect, which is the equivalent of the overdrive current in Fig. 8.12:

$$Q_{BX} = I_{BX} \tau_S$$

where I_{BX} is the excess base drive above that necessary to take the transistor to the edge of saturation. τ_S is the effective lifetime of the excess minority carriers in the base region.

The total charge required to turn the transistor ON is that required to take it to the point of saturation:

$$Q_{ON} = Q_B + Q_V = I_C \tau_{C0} + K C_{TC} \, \delta V_{CB}.$$

To turn the transistor OFF, it is necessary to extract the excess base charge, in addition to Q_{ON}.

$$Q_{OFF} = Q_B + Q_V + Q_{BX}$$
$$= I_C \tau_{C0} + KC_{TC} \, \delta V_{CB} + I_{BX} \tau_S.$$

The time constants are evaluated as follows.

$$\tau_{C0} = 1/\omega_T \quad \text{and} \quad \tau_S \simeq h_{FE(sat)} \tau_{C0}$$

where ω_T is the frequency at which $|h_{fe}| = 1$, assuming that the high-frequency asymptote falls at 20 dB/decade. The results is taken at $V_{CB} = 0$. Now

$$I_{BX} = I_B - I_C/h_{FE}$$

therefore

$$Q_{OFF} = I_C \tau_{C0} + Q_V + \tau_S(I_B - I_C/h_{FE}).$$

As an example, if $\omega_T = 25 \times 10^6$ rad/s, $h_{FE} = 50$, $h_{FE(sat)} = 25$, $C_{TC} = 12$ pF, $V_{CC} = 10$ V, and $R_L = 2$ kΩ, estimate the values of Q_{ON} and Q_{OFF} if $K = 2$.

For the purpose of this calculation, it is assumed that $V_{CE} \simeq 0$ when the transistor is ON. $I_C \simeq 10/2 = 5$ mA, $I_B = 5/25 = 0.2$ mA. $\tau_{C0} = 1/\omega_T = 0.04 \times 10^{-6}$ s.

$$Q_{ON} = (5 \times 10^3 \times 0.04) + (2 \times 12 \times 10)$$
$$= 440 \text{ pC.}$$
$$Q_{OFF} = Q_{ON} + \tau_S(I_B - I_C/h_{FE})$$
$$= 440 + (25 \times 0.04)(200 - 5000/50)$$
$$= 440 + 1(200 - 100) = 540 \text{ pC.}$$

If this transistor is in the circuit in Fig. 8.16, the value of the capacitance in the input circuit can be calculated, since it must provide the charge necessary to switch the transistor. If the input voltage changes through 10 V, and $V_{BE(sat)} = 0.7$ V, then

$$C \simeq Q_{OFF}/\delta V = 540/(10 - 0.7) = 58 \text{ pC.}$$

Assuming that the charge is supplied to, and extracted from the transistor in an exponential manner, the switching time can be calculated from the geometry of the waveshape.

The value of Q_{OFF} may be determined experimentally by a circuit similar to that in Fig. 8.16. A repetitive square wave, with a small rise- and fall-time, preferably of the order of a few nanoseconds, is applied to the input of the circuit, and the collector voltage is observed on an oscilloscope. Collector voltage waveforms, corresponding to various values of C, are shown in Fig. 8.17. The

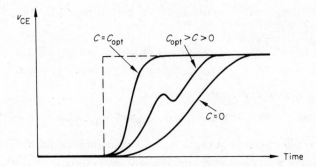

Fig. 8.17   Approximate waveforms of the collector voltage in Fig. 8.16

optimum value of capacitance, C_{opt}, is one which just eliminates the 'kinks' in
the waveform. Q_{OFF} is given by the relationship

$$Q_{OFF} = C_{opt}(V_{in} - V_{BE(sat)}).$$

Although the charge-control technique successfully describes the operation of
alloy-type transistors, it is less successful in mesa and planar devices, where
charges may be stored in parts other than the base region. For this reason the
simple mathematical model developed here should be used with care. More
extensive work on the subject of charge parameters will be found in the
references at the end of the chapter, and further study can be undertaken in the
books mentioned in the bibliography at the end of this book.

An additional method of reducing the turn-off time is to use a bias supply,
which reverse biases the emitter junction when the input voltage is zero. This
arrangement is shown in Fig. 8.18. When V_B is zero the reverse bias applied to
the base-emitter junction is approximately, neglecting the effect of the leakage
current, $-V_{BB}R_1/(R_1 + R_2)$ V. The function of the bias circuit is to provide a
transient reverse base overdrive current to discharge the base region.

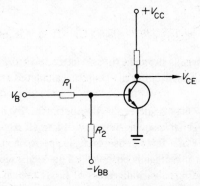

Fig. 8.18   A reverse bias applied to the base reduces the turn-off time

One method of preventing the transistor being driven too hard into its saturated state is by means of the anti-saturation circuit in Fig. 8.19. The onset of saturation occurs when the collector voltage falls below the base voltage. Diode D in Fig. 8.19 becomes forward biased after the onset of saturation but before the transistor is heavily saturated, and diverts some of the excess base drive into the collector of the transistor. Since the diode only carries the excess base current, the storage time of the diode is much less than that of an overdriven transistor, so that the turn-off time of the circuit in Fig. 8.19 is less than that of the simple invertor circuit. A diode used in this mode is sometimes described as a *clamping diode*.

Ideally, the clamping diode should have zero storage time, which is obtained in TTL gates by the use of Schottky diodes[1] (see chapter 9).

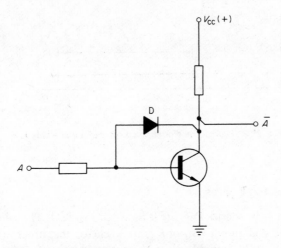

Fig. 8.19   Anti-saturation diode

8.10   Noise Immunity

The noise immunity of a logic gate is the electrical noise voltage it will withstand on any input line (or power supply line), without causing the output to register a change in voltage.

The d.c. noise margins are specified for an inverting gate in terms of its transfer characteristic (see Fig. 8.20). When the input signal is logic '0', the output voltage, V_a, corresponds to the logic '1' level, and the device operates at point X on the characteristic. The noise margin which may be allowed on the input signal in this state is NM_0. When a logic '1' is applied to the input, the device operates at point Y on the characteristic, and the d.c. noise margin is NM_1. Should the values of voltage associated with the '0' and '1' logic levels differ from those shown, then the allowable noise margins change also. Manufacturers usually quote the statistically worst value of noise margin.

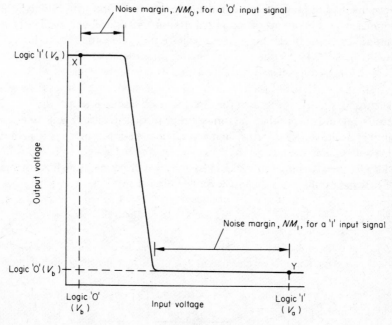

Fig. 8.20   Typical transfer characteristic of an inverting gate

8.11   Resistor logic gates

A simple positive logic resistor OR gate is shown in Fig. 8.21. The values of the input resistors, R, and output resistor R_0, are so chosen that if one or more of the input lines are at the logical '1' level, the output voltage, V_0, rises above the minimum level corresponding to logical '1'. When all inputs are at the logical '0' level, V_0 is below the logical '1' level, and is recognized as '0'.

   The AND version is obtained by connecting R_0 to a positive potential. The

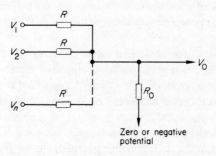

Fig. 8.21   Simple resistor logic OR gate

output potential must remain below the logical '1' level when one or more of the input lines is at '0'. The output potential rises to the '1' level only when all the inputs are at the '1' level.

8.12   Diode-resistor gates

One problem with resistor logic gates is that the inputs are permanently connected to the output through a conducting path. This has the effect of altering the output voltage when different numbers of input lines are used, but on the other hand, the resistor logic gates are cheap, and simple to construct.

A positive logic diode-resistor OR gate is shown in Fig. 8.22. With both input lines at zero potential, the net e.m.f. acting in the circuit is zero, and the output potential is zero. If input line A is connected to $+E_s$ V (logical '1'), diode A is forward biased and current flows through it, and resistor R. Since the forward p.d. across the diode is small, the output voltage, V_o, is approximately equal to $+E_s$. Under this condition the anode of diode B is at zero potential, and its cathode is at $+E_s$, i.e., it is reverse biased and no current flows through it.

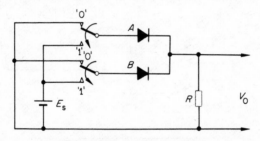

Fig. 8.22   Diode-resistor OR gate

This circuit illustrates the principal difference between resistor and diode-resistor logic. In resistor logic the value of V_o is affected by the number of input lines connected, since all those at the '0' level reduce the output voltage. In diode-resistor OR logic the diode acts as a switch, which either connects the input directly to the output when the diode is forward biased, or it isolates the input from the output when the diode is reverse biased.

When line A, in Fig. 8.22, is at '0', and line B is at the '1' level, diode B conducts while diode A is reverse biased, and the output voltage is again approximately $+E_s$ V. When both A and B lines are at the '1' level, both diodes conduct and V_o is again approximately $+E_s$ V. The truth table for the circuit is given in Table 8.1.

A positive logic diode-resistor AND gate is shown in Fig. 8.23. If one or more of the input lines are at the '0' level (zero potential), those diodes are forward biased and conduct, irrespective of the signals applied to the other lines. The p.d.

Table 8.1

Truth tables for Figs. 8.22 and 8.23

Input A	Input B	Output from Fig. 8.22	Output from Fig. 8.23
0	0	0	0
1	0	1	0
0	1	1	0
1	1	1	1

'1' = high potential
'0' = low potential

across the diodes, when conducting, depends on the current flowing and the construction of the diode. Generally the p.d. across the diode lies between 0·3 and 0·7 V, giving a potential at the output of the gate of this magnitude, which is taken as the logical '0' level.

Only when both input lines are at the '1' level does the output signal rise to logic 1, i.e., when $A = B = 1$.

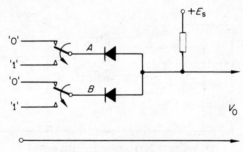

Fig. 8.23 Diode-resistor AND gate

8.13 Resistor-transistor logic (RTL)

A simple positive logic resistor-transistor NOR gate, with a fan-in of N and a fan-out of M, is shown in Fig. 8.24. The *fan-in* is the maximum number of input lines that may be connected to the gate. The OR logic function is performed by the resistor network, while the transistor provides logical inversion. The circuit works well, provided that $V_{CE(sat)}$ is less than $V_{BE(sat)}$, i.e., the collector voltage in the ON state is insufficient to turn ON any transistor connected to it. As the working temperature rises, $V_{CE(sat)}$ rises and $V_{BE(sat)}$ falls, with the result that there is an upper limit to the working temperature of the circuit. An additional drawback is that the switching speed of the circuit is relatively slow (for an electronic circuit). These limitations can be minimized by the provision of a negative bias supply to the base in the manner outlined earlier in the chapter. The design principles for the circuit shown are now outlined.

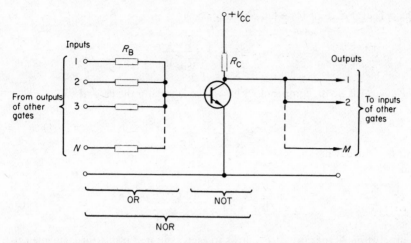

Fig. 8.24   Resistor-transistor logic (RTL) NOR gate with a fan-in of *N* and a fan-out of *M*

The worst operating conditions occur when the transistor is in the OFF state, when it must provide adequate base drive to all the connected transistors to maintain them in the ON, or saturated state. To ensure *worst-case* conditions it is assumed that all the other input lines, to the connected transistors, are supplied by transistors in the ON state, as shown in Fig. 8.25. The effects of component tolerance and supply voltage variation are considered in Example 8.1.

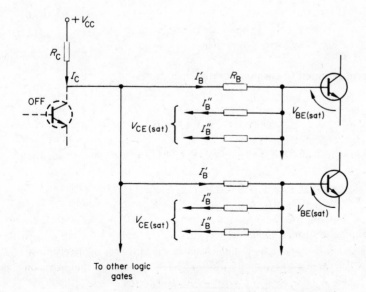

Fig. 8.25   Worse-case design procedure for the NOR gate in Fig. 8.24

The collector current for a fan-out of M is

$$I_C = \frac{V_{CC} - V_{BE(sat)}}{R_C + R_B/M}.$$

If $V_{BE(sat)}$ is small, compared with V_{CC}, as is usually the case, then

$$I_C = \frac{V_{CC}}{R_C + R_B/M}$$

and

$$I_B' = I_C/M = \frac{V_{CC}}{MR_C + R_B}. \tag{8.2}$$

A proportion I_B, of current I_B', flows to each saturated transistor connected to the gate. Since N input lines are connected to each logic element, then $(N-1)I_B''$ flows away from the base of the transistor, hence

$$I_B' = I_B + (N-1)I_B''. \tag{8.3}$$

Current I_B'' is given by

$$I_B'' = \frac{V_{BE(sat)} - V_{CE(sat)}}{R_B}.$$

Allowing for worst-case values, $V_{CE(sat)}$ is assumed to be zero, giving

$$I_B'' = \frac{V_{BE(sat)}}{R_B}. \tag{8.4}$$

The base current required to ensure saturated operation is

$$I_B = \frac{I_C}{h_{FE(sat)}} \cong \frac{V_{CC}}{R_C h_{FE(sat)}}. \tag{8.5}$$

Substituting eqs. (8.2), (8.4), and (8.5) into eq. (8.3) gives

$$\frac{V_{CC}}{MR_C + R_B} = \frac{V_{CC}}{R_C h_{FE(sat)}} + \frac{(N-1)V_{BE(sat)}}{R_B}.$$

Solving this expression for M yields

$$M = \frac{h_{FE(sat)}}{1 + \frac{(N-1)V_{BE(sat)}h_{FE(sat)}}{V_{CC}R_B/R_C}} - \frac{R_B}{R_C}.$$

To obtain a large fan-out it is desirable that:

(a) $h_{FE(sat)}$ should be large.
(b) V_{CC} should be large.
(c) R_B/R_C should be small.
(d) N should be small.
(e) $V_{BE(sat)}$ should be small.

It is clear, from condition (d), that there is a trade-off between fan-in and fan-out, and one can be improved at the expense of the other. The speed of operation of the circuit in Fig. 8.24 is improved by providing a negative bias rail, but this has a detrimental effect on the fan-out capability, since the bias current has to flow through the collector and base resistors. This current results in an increased p.d. in these resistors, and reduces the available base drive.

To avoid the possibility of noise signals affecting the performance of NOR circuits, it is advisable to connect unused inputs to the zero potential line.

Example 8.1: Design a resistor-transistor NOR gate similar to Fig. 8.24 using a 20 V supply. When the collector current is 10 mA, $h_{FE(sat)}$ is 10 and $V_{BE(sat)}$ is 0·7 V.

Show by means of a graph the effect of component tolerance on the fan-in and fan-out.

Solution: Assuming that all the parameters and voltage levels have constant values, values of M, for various values of N, are calculated for various values of the ratio R_B/R_C. These are shown in graphical form in Fig. 8.26(a). The fan capability is seen to be reduced for the higher and lower values of the ratio R_B/R_C, compared with the case when the ratio is unity. The optimum ratio is dependent on the fan-in or fan-out required, and the ratio will clearly be less than 5 and greater than 0·1.

Taking a ratio $R_B/R_C = 1$, it is permissible to use $R_B = R_C = 2$ kΩ to give a saturation current of 20 mA with $V_{CC} = 20$ V. The fan-out capability is clearly a function of the fan-in, and, if five input lines are used, the fan-out must be limited to three.

The effect of changes in component values and voltage variations are next considered, for the chosen ratio of R_B/R_C. These are plotted, in Fig. 8.26(b), for a ratio of $R_B/R_C = 1$, for worst case 10% and 20% variations. These variations reduce the fan capability even further. In the 20% case, if $N = 5$ the fan-out is reduced to two.

It has clearly been demonstrated, in Example 8.1. that there is a relationship between the fan-in of the driven stage and the fan-out of the driver stage. As an added complication, the fan-out of any stage affects its own fan-in, and vice versa. This is inherent in semiconductor devices as the input and output circuits are electrically connected by the internal conduction paths. To overcome these problems when designing logic networks, manufacturers normally provide

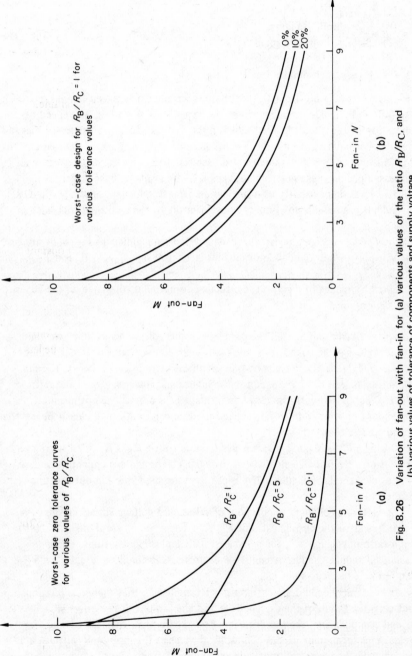

Fig. 8.26   Variation of fan-out with fan-in for (a) various values of the ratio R_B/R_C, and (b) various values of tolerance of components and supply voltage

information relating the fan-in and fan-out under specific operating conditions

Transistor manufacture has not yet reached the stage where each transistor of a given type has the same parameter values. For example, $V_{BE(sat)}$ can vary widely between transistors of the same type, with the result that if a number of driven gates are supplied from one driver gate, the transistor which saturates first tends to pass more base current than the others, possibly preventing them from saturating. This is known as *current hogging*. The effects of this can be minimized by using a relatively high value of series resistor in the base circuit, making all the transistor input resistances similar.

8.14   Resistor-transistor *S-R* flip-flop

It was shown, in chapter 5, that two NOR gates can be used to construct an *S-R* memory. The resistor-transistor *S-R* version is shown in Fig. 8.27, comprising two NOR gates with direct feedback between them. A bias supply is used to reduce the turn-off time, and the feedback resistors are shunted by capacitors to reduce the switching time. The circuit in Fig. 8.27 is often referred to as a *S-R bistable* circuit since it has two stable states, corresponding to one or other of the transistors being in the ON state at any one time. In practice both transistors are ON simultaneously during the very short time that the bistable is changing state.

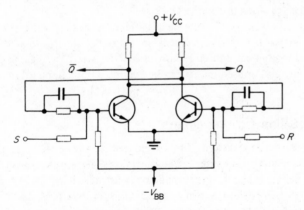

Fig. 8.27   The circuit diagram of an *S-R* flip-flop using resistor-transistor logic

8.15   Diode-transistor logic (DTL)

The positive logic NOR gate, in Fig. 8.28(a), comprises a diode-resistor OR circuit followed by a transistor NOT gate, the NOR function being generated overall. The negative bias rail is used to reduce the turn-off time of the gate.

A positive logic diode-transistor NAND gate is shown in Fig. 8.28(b), comprising a diode-resistor AND gate followed by an inversion stage to give the

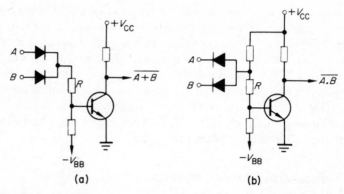

Fig. 8.28   Diode-transistor logic (DTL) (a) NOR gate, and (b) NAND gate

NAND connective. In both circuits resistor R can be shunted by a capacitor to reduce the turn-on time of the gate. Anti-saturation or clamping diodes can be used in conjunction with these circuits.

The circuit in Fig. 8.28(a) is sometimes referred to as current sourcing logic, since the gate acts as a current source to the inputs of the following stages when the transistor is in the OFF state. Figure 8.28(b) is also known as *current sinking* logic, since the transistor must absorb or 'sink' the current flowing in the diodes in connected gates when it is in the ON state.

To avoid the possibility of noise signals affecting the performance of these circuits, it is advisable to connect unused inputs on NOR gates to a 'low' (logic '0') signal and to connect unused inputs on NAND gates to a 'high' (logic '1') signal. Alternatively, unused inputs can be connected or 'strapped' to used inputs.

8.16   The Schmitt trigger circuit

There is a need, in many systems, for a circuit which gives a rapid transition from one voltage level to another, when the input signal varies slowly. One circuit is the Schmitt trigger, one form being constructed using a high gain differential input amplifier (known as an *operational amplifier*). The basic circuit is illustrated in Fig. 8.29(a). In this circuit, the input signal is applied to the *inverting input* (marked with a '−' sign) of the amplifier, and a reference signal is applied to the noninverting input (marked with a '+' sign) via R_1. The characteristic of the circuit is shown in Fig. 8.29(b) and, for input voltages less than V_1, the output voltage has a positive polarity and is equal to V_H. When the input voltage exceeds V_1, the output voltage abruptly changes to V_L (which has a negative polarity) and remains at this level so long as the input voltage is greater than V_1.

The output voltage returns to V_H gain when the input voltage is reduced to V_2. The triggering voltages, V_1 and V_2, have different values, and the difference

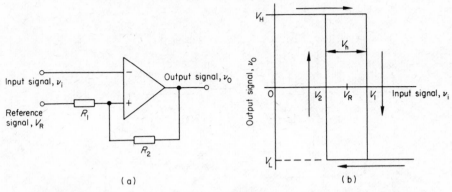

Fig. 8.29 (a) and (b)   One form of Schmitt trigger circuit using an operational amplifier

between their values is known as the *hysteresis* voltage, V_h, of the circuit. The hysteresis effect is particularly valuable in electrically noisy environments, since it provides the circuit with a degree of noise immunity.

An analysis of the circuit in Fig. 8.29 is given in the literature,[5] and a summary of the results is given here. If the magnitudes of V_H and V_L are the same, and have the value V_0, then

$$V_1 = (R_2 V_R + R_1 V_0)/(R_1 + R_2)$$
$$V_2 = (R_2 V_R - R_1 V_0)/(R_1 + R_2)$$
$$V_h = V_1 - V_2 = 2R_1 V_0/(R_1 + R_2)$$

8.17   The monostable multivibrator

The monostable multivibrator is a circuit which has one stable state and one quasi-stable state. The circuit is triggered into its quasi-stable state by means of an impulsive signal, in which it remains for a period of time determined by the circuit time constants, after which it returns to its stable state.

A typical circuit diagram of a monostable multivibrator is shown in Fig. 8.30(a). In its stable state, the current flowing through resistor R saturates TR2 and, at the same time, TR1 is held in a cut-off state by the bias supply V_{BB}. In this condition, capacitor C is charged to approximately V_{CC} with the polarity shown. The circuit is switched into its quasi-stable state either by turning TR1 ON, or by turning TR2 OFF. Transistor TR1 can be turned ON by applying a positive-going pulse to trigger input A, and TR2 can be turned OFF by applying a negative-going pulse to input B; the waveforms in Fig. 8.30(b) correspond to the former mode of triggering. The duration of the trigger pulse should be short when compared with the length of the quasi-stable period of the circuit.

When TR1 is triggered into conduction by the application of a positive-going pulse to input A, the left-hand plate of capacitor C is instantaneously earthed. This applies a potential of $-V_{CC}$ to the base of TR2, cutting this transistor OFF. The base voltage of TR2 recovers from this voltage towards a potential of

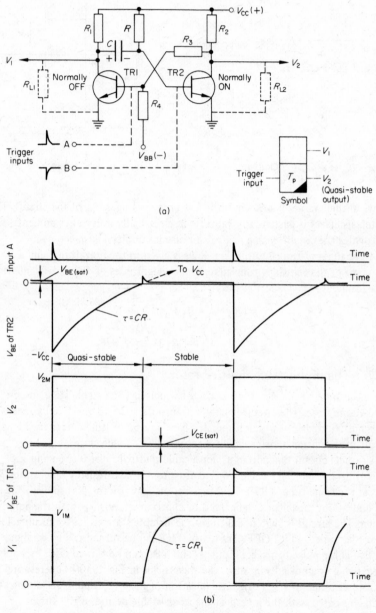

Fig. 8.30   A monostable multivibrator

$+V_{CC}$ along an exponential curve with a time constant of RC, as capacitor C charges through resistor R. The instantaneous value of the base voltage is given by the expression

$$v_{B2} = -V_{CC} + 2V_{CC}(1 - e^{-t/RC})$$

Assuming that TR2 begins to conduct when v_{B2} reaches a potential of V_{B2}, then the expression for the period, T_p, of the pulse at the collector of TR2 is

$$T_p = RC \, ln \left[2 / \left(1 - \frac{V_{B2}}{V_{CC}} \right) \right] \tag{8.6}$$

A simpler expression is obtained if it can be assumed that TR2 begins to conduct when its base voltage reaches zero, giving

$$T_p = RC \, ln \, 2 = 0 \cdot 693 \, RC \simeq 0 \cdot 7 \, RC \tag{8.7}$$

After this period of time, TR2 returns to the ON state and, due to the bias resistor chain $R_3 R_4$, causes TR1 to be turned OFF. The recovery of the collector voltage of TR1 to its full value of V_{CC} is delayed by the rate at which capacitor C is charged through resistor R_1. Hence the collector voltage of TR1 rises along an exponential curve of time constant CR_1; the transient period of this curve is about $5CR_1$ seconds. A simple modification which allows the slow rise in the collector voltage of TR1 to be overcome is described in section 8.18.

8.18   The astable multivibrator

An astable multivibrator or *free-running* multivibrator, is a circuit which has two quasi-stable states, one following the other in succession. A popular version of the circuit is shown in Fig. 8.31.

Complementary outputs are available from the collectors of the two transistors, and it can be seen that each half of the astable multivibrator is generally similar to the left-hand half of the monostable multivibrator in Fig. 8.30. Assuming that each transistor begins to conduct when its base voltage rises to zero, then from eq. (8.7)

$$T_{p1} \simeq 0 \cdot 7 \, R_4 C_2$$
$$T_{p2} \simeq 0 \cdot 7 \, R_3 C_1$$

The periodic time, T, of the complete waveform is

$$T = T_{p1} + T_{p2} \simeq 0 \cdot 7 (R_4 C_2 + R_3 C_1)$$

If, as is often the case, a common value R is chosen for R_3 and for R_4, and a common value C is selected for the timing capacitors then

$$T \simeq 1 \cdot 4 \, RC$$

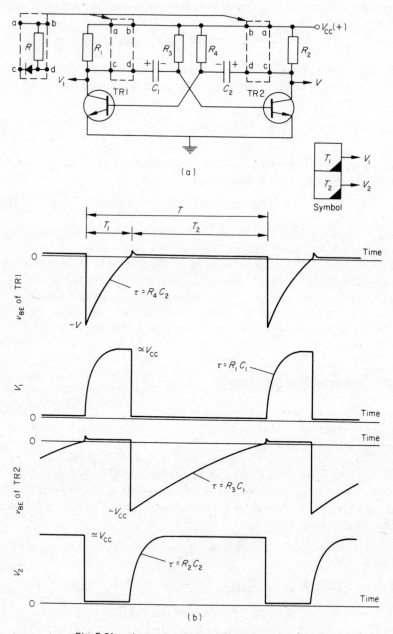

(a)

(b)

Fig. 8.31   A popular astable multivibrator circuit

The output waveforms from this circuit do not have a good square waveshape; this is due to the time taken for the timing capacitors to fully charge via the collector resistors (see also Fig. 8.30(b)). The waveforms are improved by the addition of the diode-resistor combination shown in the inset to Fig. 8.31(a). The diode effectively isolates the capacitor charging current from the collector of the associated transistor, thereby allowing the collector voltage to rise quickly when the transistor is turned OFF.

8.19   Field-effect transistors

As the name of these devices suggests, an electrical field (rather than a current) is used to control the flow of current through them. There are two principal types of FET, namely:

1. The junction-gate FET (the JUGFET)
2. The insulated-gate FET (the IGFET or MOSFET)

The principal area of application of JUGFETs is in linear electronics, which is not of direct concern to us here. Their construction and operation is described in the literature.[1,5,6]

8.19.1   Insulated-gate field-effect transistors

The insulated-gate field-effect transistor forms the basis of many logic devices used in integrated circuit form (see also chapter 9). Without it, the development of battery-operated pocket calculators would have been impossible.

A section through a basic element, the p-channel IGFET or MOSFET, is shown in Fig. 8.32(a). The source and the drain act as the respective electrodes at which the mobile charge carriers are injected into and extracted from the device. For reasons given later, only one type of charge carrier is involved in conveying current between the source and the drain which, in the case considered here, is the hole; since only one type of charge carrier is involved, these devices are sometimes described as *unipolar transistors*. Since, in Fig. 8.32(a), holes are used as charge carriers, the source electrode is connected to the positive pole of the supply, and the drain to the negative pole. The gate region of the device provides a method of controlling the value of the drain current, and is insulated from the semiconducting substrate by a thin silicon oxide layer. The popular name of MOSFET applied to this device is derived from the gate-to-substrate structure (*M*etal-*O*xide-*S*emiconductor *FET*).

In the basic p-channel device, two p^+-regions are diffused into an n-type substrate. Since the drain region is connected to the negative pole of the supply, a depletion region forms at the drain junction and, initially, the drain current is zero. The substrate is manufactured from low conductivity n-type silicon, so that only a relatively few mobile electrons are available. At normal values of ambient temperature, electrons and holes are continuously, but randomly, generated in the body of the substrate. The application of a negative potential to

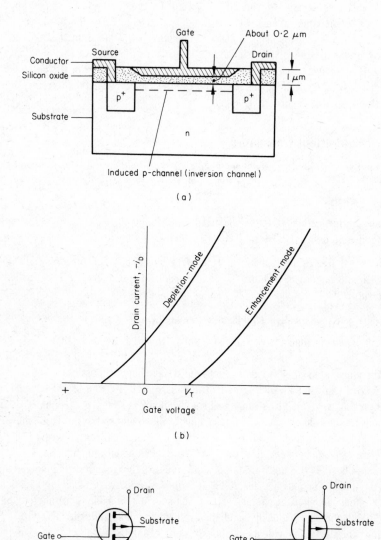

Fig. 8.32   (a) A section through a p-channel MOSFET, (b) typical transfer characteristics for p-channel devices, (c) symbol for an enhancement-mode device, and (d) symbol for a depletion-mode device

the gate electrode of the MOSFET causes the holes to be attracted to the metal-to-semiconductor interface beneath the gate electrode. At a gate potential known as the *threshold voltage, V_T*, a sufficient number of the thermally generated holes have collected in sufficient numbers on the underside of the oxide layer to form a conducting *channel* (known as an *induced channel* or as an *inversion channel*) between the source and the drain. Typical values of V_T lie in the range -2 to -5 V; devices which are manufactured using the silicon-gate process[1] have threshold voltages in the range $-1 \cdot 5$ to $-2 \cdot 0$ V. Increasing the value of the gate voltage, i.e., making it more negative with respect to the source electrode, attracts more holes to the interface, thereby increasing both the conductivity of the channel and the value of the drain current. The depth of the inversion channel formed in this way is typically a few angstrom units $(1 \text{ Å} = 10^{-10} \text{ m})$.

Since the value of the drain current is zero with zero gate voltage and, for gate voltages greater than V_T, is increased or enhanced, this type of device is known as an *enhancement-mode* MOSFET. A typical transfer characteristic or mutual characteristic for a p-channel enhancement-mode MOSFET is illustrated in Fig. 8.32(b). The circuit symbol for a device of this kind is shown in Fig. 8.32(c); the broken line between the source and drain electrodes indicates that, for zero gate voltage, no current flows between the two electrodes. In the symbol, the gate electrode is separated from the source-to-drain channel to indicate the electrical isolation between the two. The arrow pointing from the central part of the channel indicates the nature of the junction between the channel and the substrate; it points from the p-region (the channel) to the n-region (the substrate), i.e., it follows the diode convention.

Enhancement-mode MOSFETs can also be used to replace resistors in logic invertor circuits with some advantage since, for a given equivalent ohmic value, they require a much smaller area of the integrated circuit than does the resistor. This effectively either reduces the size (and cost) of a given logic gate, or allows a more complex logic circuit to be manufactured in the same physical space. An example of a MOSFET used in this way is given in section 8.20.

Another type of MOSFET, the *depletion-mode* MOSFET, has the type of mutual characteristic which is also shown in Fig. 8.32(b); the circuit symbol for a p-channel depletion-mode MOSFET is given in Fig. 8.32(d). This type of device, which is less common than the enhancement-mode device, has an *initial channel* of p-type semiconductor diffused into the structure between the source and the drain electrodes. This allows current flow to take place between these regions when the gate voltage is zero. This fact is illustrated in the circuit symbol by the solid line linking the source to the drain. The application of a negative potential to the gate electrode attracts thermally generated holes to the conducting channel, thereby increasing the drain current as before. However, in this type of device, the application of a positive potential to the gate region causes the holes in the initial channel to be repelled away from the oxide-semiconductor interface; this reduces the channel conductivity and the value of the drain current. Hence, in such an element it is possible to reduce or deplete the drain current below the zero bias value. Field-effect devices which have a finite drain current for zero gate voltage, and whose value can be reduced or depleted by altering the value of the gate voltage are known as *depletion-mode* devices.

N-channel MOSFETs are also manufactured but, due to a number of limitations,[1,5,6] are not so widely used in logic circuits as are p-channel devices. However, n-channel devices are used in conjunction with p-channel devices in the form of *complementary MOS* logic circuits (CMOS or COSMOS), one form of circuit being discussed in section 8.20 and, in chapter 9, a more general treatment is given.

Since the effective input circuit of the MOSFET is a capacitance shunted by a high value of resistance (typically 10^{12} Ω), the input time constant is such that their frequency response is not so good as that of bipolar transistors. However, the advantages of MOSFET circuits are such that they are widely used in logic circuits; these advantages include small size, low power consumption, high noise immunity and a tolerance to supply voltage variations.

8.20   MOS NOT gates

Two types of MOS NOT gates are illustrated in Fig. 8.33. In the p-channel gate in Fig. 8.33(a), TR2 is so biased and constructed that it replaces the loading resistor in a conventional NOT gate. This circuit operates with a supply voltage of about −20 V, and uses the negative logic notation; typical voltage levels are

$$\text{logic '1': } -11 \text{ V to } -14 \text{ V}$$
$$\text{logic '0': } -2 \text{ V to } -3 \text{ V}$$

The complementary MOS gate in diagram (b) contains both p- and n-channel devices, both being switched by the input signal. That is, when TR3 is ON then TR4 is OFF and vice versa. For this reason, TR4 is described as an *active load*. Consequently, the mean power consumption of the gate is only a few microwatts. CMOS logic uses a positive supply potential and the positive logic notation is used to describe its operation; the supply voltage can be in the range 3–15 V. When the input signal is logic '0', TR3 is OFF and TR4 is ON; in this state, the output line is coupled to the positive supply line. When the input signal is logic '1', TR3 is ON and TR4 is OFF; the output line is then coupled to ground. From the foregoing, the output resistance of the circuit is low (typically 500 Ω) for either level of output signal. With a 5 V supply, typical output voltages from a CMOS gate of the type shown are

$$\text{logic '1': } 4.95 \text{ V}$$
$$\text{logic '0': } 0.05 \text{ V}$$

This type of gate has a large noise margin, and is typically $0.45\ V_{\text{S}}$, i.e., 2.25 V with a 5 V supply. Since the input impedance of this gate is very high, the input

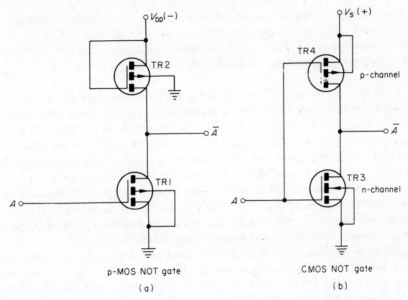

Fig. 8.33   (a) A p-channel MOS NOT gate, and (b) a complementary MOS NOT gate

current has a very small value (typically 10 nA), giving each gate a fan-out in excess of 1000 at low operating frequencies (that is, below about 10 kHz).

8.21   Noise in logic circuits

Noise is defined as all spurious signals, random and otherwise, that are not part of the input information. Noise is generally regarded as being either *narrowband* or *broadband*. Narrowband noise is low frequency noise, typically in the range 10 Hz to 1 kHz, and is often harmonically related to the power supply frequency. Broadband noise comprises signals above 1 kHz, and is generally produced by thermal effects in resistors, and other effects in semiconductors and thermionic devices. In logic networks, the effect of many gates switching simultaneously gives rise to electrical noise. It is generally desirable for logic gates to possess a high degree of *noise immunity* to prevent inadvertent operation by noise signals.

Noise can be divided into two types, *common mode* and *normal mode*. Common mode signals appear between each conductor and earth, and many special circuits have been designed to have a high common mode rejection. Normal mode noise appears between conductors. There are five principal causes of noise in logic circuits:

1. Electrostatic noise, due to the effects of circuit capacitance.
2. Electromagnetic noise, due to magnetic coupling between circuits.

3. Thermal noise, due to the electrothermal properties of the materials and devices used in the circuits.
4. Noise due to the leakage of current across insulation.
5. Noise resulting from the variation in the power supply voltage.

The resistor-transistor NOR gate, in Fig. 8.24, possesses only a low degree of noise immunity, since a noise voltage in excess of $V_{BE(sat)}$, at the base of the transistor, will switch it to the ON state. The noise immunity of Fig. 8.24 is improved by using a relatively high value of resistance for each of the input resistors, with consequent reduction in switching speed and fan-out. If 'speed-up' capacitors are used in parallel with the input resistors, the noise immunity is reduced, since a 'noise' voltage spike of magnitude $V_{BE(sat)}$ at the collector of any transistor will be transmitted directly to the base of the following transistor. In general the use of speed-up capacitors should be avoided where possible, as these transmit broadband noise without attenuation.

The bias rail V_{BB} in Fig. 8.28 has the added advantage of increasing the noise rejection of those gates, as well as reducing the turn-off times. It is necessary for the noise voltage, at the input to the gate, to exceed the reverse bias applied to the transistor before the gate can be turned ON.

The forward conducting properties of diodes can also be utilized to increase noise rejection, as well as simplifying circuits. An example is shown in Fig. 8.34. By replacing resistor R, in Fig. 8.28(b), with the diodes D1 and D2 in Fig. 8.34,

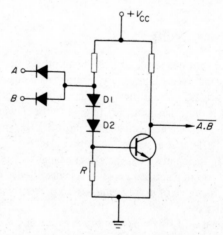

Fig. 8.34   Diodes D1 and D2 improve the noise rejection of the gate

the noise voltage at the input has to exceed the forward voltage drop of D1 and D2 before the transistor base potential begins to change. More diodes can be used if necessary. With this connection it is not necessary to use a negative bias supply since, with inputs A and B in the '0' state, the transistor base voltage is

practically zero due to the forward conducting properties of D1 and D2.
Resistor R, in Fig. 8.34, provides a low impedance path for the charge carriers in the base region of the transistor when it is turned OFF. This resistor can be omitted if D1 and D2 are slow-recovery diodes, i.e., the transistor is capable of turning OFF at a faster rate than the diodes. In this event the charge carriers in the transistor base region, when the transistor is turned OFF, return to ground via D1, D2, and the input diodes. The latter arrangement is convenient in circuits using discrete components, but presents problems in monolithic circuit elements, since it is difficult to manufacture devices with different reverse recovery times in the same slice of silicon. Noise rejection can be further improved by replacing D1 and D2 by a Zener diode.

It was stated earlier in this chapter that, in RTL, a high value of input resistance is desirable to minimize the effects of 'current hogging'. A high value of input resistance also increases the noise immunity. For a given current, it increases the p.d. between the collector of the driving transistor and the base of the driven transistor to drive it into the ON state. On the other hand, high values of input resistance reduce the switching speed of the gate due to the longer propagation time.

Problems

8.1. Given that the small-signal forward current transfer ratio h_{fe} of a transistor, at frequency ω, is given by

$$h_{fe} = h_{FE}/(1 + j\omega/\omega_\beta)$$

show that the expression for the collector current change, for a step change I_B in the base current, is

$$I_C = h_{FE}I_B(1 - e^{-t/T})$$

where

$$T = 1/\omega_\beta.$$

8.2. In a resistor-transistor NOT gate, a transistor with an h_{FE} of 20 is used. The collector saturation current is 6 mA, and the collector voltage is 'caught' at 2·5 V; under these conditions the base current is 0·3 mA. If a current of 0·1 mA is extracted from the base circuit by a constant current source, for 'speed-up' purposes, calculate the value of the resistor used in the input circuit.

8.3. Show that the NOR logic function is generated by connecting together the outputs of a number of NOT gates.

8.4. Describe the operation of a resistor-transistor NOR element.

8.5. Compare diode-transistor logic with resistor-transistor logic.

8.6. Draw a circuit diagram of diode-transistor (a) AND gate, and (b) OR gate. State whether positive or negative logic is used, and explain the operation of the circuits. What logic functions are performed in each case if the logic levels are inverted?

References

1. MORRIS, N. M., *Electronic Devices*, Macmillan.
2. MORRIS, N. M., *Control Engineering*, McGraw-Hill.
3. BEAUFOY, R. and J. J. SPARKES, 'The junction transistor as a charge controlled device', *ATE Journal*, **13**, 310, 1957
4. BEAUFOY, R., 'Transistor switching circuit design using the charge control parameters', *Proc. Inst. Elect. Engrs.*, **106 B**, Suppl. No. 17, 1092, 1959.
5. MILLMAN, J., and C. C. HALKIAS, *Integrated Electronics*, McGraw-Hill.
6. MORRIS, N. M., *Advanced Industrial Electronics*, McGraw-Hill.

9. Integrated electronic circuits

Progress in electronics has resulted in the miniaturization of equipment with the benefit of space-saving and the by-products of more economic and reliable components. The art of miniaturization has led to the development of integrated circuits. At their present state of development, integrated circuits are broadly divided into three groups: (a) film circuits, (b) monolithic integrated circuits, (c) hybrid circuits.

9.1 Film circuits

In the film-integrated circuit passive elements are manufactured by depositing films of conducting and non-conducting materials on an insulating or passive substrate. Materials used for the substrate include borosilicate glass, and ceramic materials having a high alumina content.

Resistors are in the form of a film of conducting material, nickel–chromium alloys being typical, having a thickness of a few millionths of an inch. During manufacture it is convenient to deposit all the resistors in one operation, leading to a uniform depth of conductor material. A feature of components manufactured by film techniques is that they can be accurately adjusted in value during manufacture by cutting away part of the component. The resistant (R) between opposite faces of a bar of conducing material is given by

$$R = \rho l / wd \quad \Omega$$

where ρ = resistivity of the bar in Ω.m,
  l = length of the bar in m,
  w = width of the bar in m,
  d = depth of conducting material in m.

If $w = l$, i.e., the bar is square in plan view, then $R = \rho/d$. This expression is independent of the physical sizes of both w and l. In any given manufacturing process depth d is constant, and the ratio ρ/d is usually defined as the *sheet resistance* in ohm per square, and is assigned the symbol p. Thus the resistance of a conducting bar of length l and width w is

$$R = pl/w \quad \Omega. \tag{9.1}$$

For a given sheet resistance the ratio of length to width can be calculated for any value of resistance. If $p = 500$ ohm per square, a 5 kΩ resistor requires the ratio $l/w = 10$.

The power dissipated per unit area of the substrate is limited by heating effects. If the maximum permissible power dissipation per unit area is W_m watt, then the actual power dissipated, W, is

$$W = W_m \times \text{area} = W_m wl$$

or

$$wl = W/W_m. \tag{9.2}$$

With given values of R, p, and W_m the geometry of the film resistor can be calculated. The maximum power dissipation depends to a great extent upon the substrate material, 3 W/cm^2 being acceptable on glass substrates, but this is increased to $15\cdot5$ W/cm^2 on glazed ceramic materials.

Example 9.1: Calculate the dimensions of a film resistor of value 5 kΩ given that $p = 500$ Ω/square, $W_m = 2$ watt/cm^2, and that the power dissipation of the resistor is to be $0\cdot1$ watt.

Solution: From eq. (9.2)

$$W/W_m = 0\cdot1/2 = 0\cdot05 = wl \tag{9.3}$$

From eq. (9.1)

$$l = Rw/p.$$

Substituting this value in eq. (9.3) gives

$$Rw^2/p = 0\cdot05$$

or

$$
\begin{aligned}
w^2 &= 0\cdot05\, p/R \\
&= 0\cdot05 \times 500/5000 = 0\cdot005.
\end{aligned}
$$

therefore

$$w = 0\cdot0707 \text{ cm.}$$

$$l = 0\cdot05/w = 0\cdot707 \text{ cm.}$$

Capacitors for film-integrated circuits are often manufactured in the parallel plate form, comprising successive layers of conducting and insulating material.

The range of values of capacitance obtainable is restricted by the substrate area available and the thickness of the insulating material that may be used, the latter depending on the breakdown strength of the insulating material. The maximum value of capacitance at present practicable is a few thousand picofarads. Above this value, capacitors are attached externally as discrete components.

Inductances of a few microhenry can be manufactured by depositing conducting material in a spiral form, but these occupy a large area on the substrate and have a low Q-factor Whenever possible the use of inductors is avoided by re-designing the circuit to make use of active components. If this is not possible, inductors may be added as discrete components external to the film circuit.

Transistors can be manufactured in small quantities using film technology in the form of MOSFETs. Active elements can be added as discrete components to the film circuit. The transistors used are in the form of a small piece or *chip* of semiconductor material. The chip is 'flipped' or turned over to enable connections to be made to it, as shown in Fig. 9.1, the connections being made by ultrasonic or thermo-compression bonding. This is known as the *flip-chip* method of mounting active devices. Since active components are added to passive film circuits, the circuit designer can initially test out all his circuits using a discrete component design and worst-case design techniques. The design can then be transferred directly to a film circuit. Film circuits are best utilized where the ratio of active to passive elements is low. For small batch production of special circuits, the film circuit is cheaper than the monolithic circuit, since these require high production rates to justify their economics.

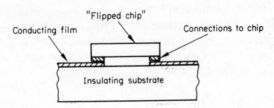

Fig. 9.1   'Flip-chip' method of mounting an active device on a passive substrate

Film circuits are subdivided into thin- and thick-film circuits. The division is one of relative magnitude, since both are 'thin' by normal standards. In *thin-film* circuits the conductor depth is in the range $10^{-6}-10^{-4}$ in., while *thick-film* conductors have depths of about $10^{-4}-10^{-2}$ in. The type of film depends to a large extent on the material used and the method of manufacture. The performance of both types is similar, although it is sometimes claimed that thick-film circuits are more rugged and cheaper than thin-film circuits.

The sheet resistance and range of resistance values available differs with the type of film used. Thin-film methods give sheet resistances between 10 Ω/square and 1 kΩ/square, with a range of resistance values between 10 Ω and 100 kΩ.

Thick-film sheet resistances lie between 100 Ω/square and 50 kΩ/square, with resistance values between 20 Ω and 1 MΩ. Capacitors in thin-film circuits have values between 100 and 10,000 pF, while those in thick film circuits lie between 20 and 5000 pF. It is often more convenient to add large capacitors as discrete components to the film circuit.

9.2 Monolithic integrated circuits

Monolithic or semiconductor integrated circuits are manufactured in a silicon substrate, both passive and active elements being made in the small 'chip' of about 1 mm^2. Resistors, capacitors, and transistors are relatively easy to manufacture, but at the present time no suitable replacement for inductors has been found. As in the case of the film circuit, it may be necessary to re-design the circuit if inductors are required.

In order to manufacture a range of components, several p-n junctions have to be developed in the semiconductor chip. Two methods are commonly used, epitaxial deposition and diffusion.

In the process of *epitaxial deposition*, layers are formed on the surface of a silicon chip by raising the chip to a high temperature and passing a gas containing special compounds over it. The thin layer so formed has the same crystal structure as the silicon on which it is formed, having a higher resistivity than the chip. Resistors, capacitors, diodes, and transistors are formed in the epitaxial layer by the process of *diffusion*.

Junctions are formed by diffusion as follows. The silicon chip is heated in a steam atmosphere at a high temperature, allowing the surface to oxidize. 'Windows' are then cut in the skin of SiO$_2$ by a photo-engraving process. The slice is again raised to a high temperature in a furnace, and gases containing dopants which result in a crystal structure of the opposite type to that of the substrate, diffuse into the substrate. This process is illustrated in Fig. 9.2. Transistors are formed by a process of multiple diffusion.

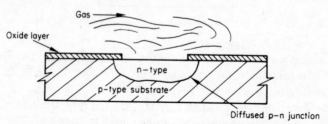

Fig. 9.2   The process of manufacturing a diffused p-n junction

A monolithic integrated resistor is shown in Fig. 9.3 together with its approximate equivalent circuit. The resistive element between A and B is the collector region of a p-n-p junction transistor. The resistor is isolated from the

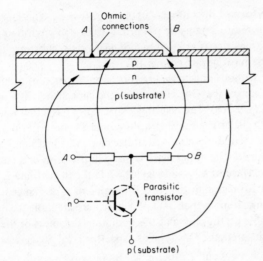

Fig. 9.3   A monolithic integrated resistor in the form of the collector region of a transistor formed in the substrate

substrate by making the n-region positive with respect to the resistive p-region. A resistor tends to take up a bigger substrate area than a transistor, and for this reason pinch-effect resistors are often preferred in monolithic circuits. The *parasitic transistor* between the two p-regions can cause feedback effects to other components in the integrated circuit. This is one of the problems associated with this type of circuit. Due to the difficulties in accurately controlling the diffusion process, the tolerance of the resistance values developed by this technique is very wide, typically ±20%.

A diode is formed using the collector junction of the transistor, shown in Fig. 9.4. The diode is isolated from the substrate by making the substrate negative with respect to the point *C*.

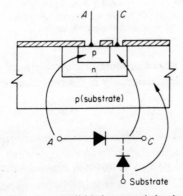

Fig. 9.4   A monolithic integrated circuit diode

Transistors are formed by a process of three diffusions. The useful n-p-n
transistor is connected to a parasitic p-n-p transistor, shown dotted in Fig.
9.5(a). The n-p-n transistor is isolated by connecting the substrate to the point in
the circuit with the most negative potential.

There are two methods in general use for the production of monolithic
integrated circuit capacitors. One is to utilize the capacitance of a reverse biased
diode, shown in Fig. 9.5(b). The resulting capacitance has a low value and is
dependent on both the applied reverse voltage and the impurity doping of the
diode regions. The second method is to utilize the oxide layer on the surface of
the chip as a dielectric. The upper plate is made in the form of a metallic film
deposited on the surface of the oxide layer, while the lower plate is a suitably
doped semiconductor region in the chip. The value of capacitance normally
obtainable is not much more than 100 pF since there are limits to both the
surface area available on the chip and the minimum thickness of the oxide layer
due to its dielectric strength. This type of capacitor is known as a MOS
capacitor.

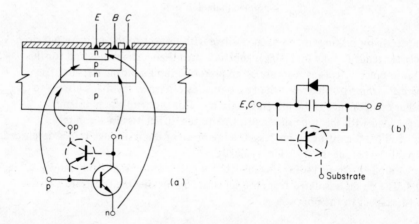

Fig. 9.5   (a) An n-p-n transistor is produced by multiple diffusion in the substrate, but is
associated with the parasitic transistor shown dotted. (b) A capacitor can be constructed by
reconnecting the transistor in (a)

Inductors cannot normally be produced by conventional semiconductor
techniques and must be added externally as discrete components.

9.2.1   Field-effect transistors

MOS devices can readily be manufactured in monolithic form, the construction
of a p-channel MOSFET having already been illustrated in Fig. 8.32(a).
Advantages of MOSFETs over bipolar junction devices include the fact that they
require as little as one-fifth of the area required by an equivalent bipolar circuit
on a semiconductor chip, and that many of the problems associated with parasitic
diodes and transistors in bipolar circuits are not present in MOS circuits.

9.2.2   Multiple emitter transistor

Monolithic production techniques have led to the manufacture of many unique devices. One of the most useful is the multiple emitter transistor (MET) illustrated in Fig. 9.6(a) which is widely used in TTL circuits. Its operation is described with reference to Figs. 9.6(b) and (c).

Inputs A, B, and C are connected to the collectors of other transistors which are either at zero potential (saturated), or at logic '1'. Figure 9.6(c) shows the simple equivalent circuit of the MET, and its similarity to the input circuit of the DTL gate in Fig. 8.34 should be noted.

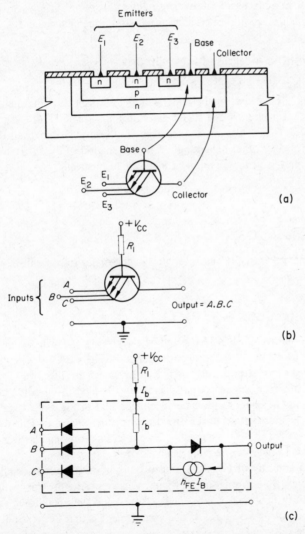

Fig. 9.6   (a) The construction and circuit symbol of the MET. In (b) the usual circuit configuration is shown together with the approximate equivalent circuit in (c)

With all inputs in the '1' state, all the emitter junctions of the MET are reverse biased. In this event the collector potential of the MET is 'high' since the collector is connected via R_1, r_b, and the forward biased collector diode. If any of the inputs are taken to '0', the base current is diverted to those emitters and the MET saturates. Since the emitter-collector potential of a saturated transistor is very small, the output voltage level falls to logical '0'. In the mode described, the MET performs the AND function.

A feature of this device is that the base current is similar in both the ON and OFF states (assuming that the output is connected to a transistor), since one or other of the MET junctions are forward biased at all times. Switching the transistor is then a matter of re-distributing the base current between the junctions, which is accomplished in approximately one nanosecond.

9.3   Monolithic integrated logic circuits

In the early days of the development of monolithic circuits, an attempt was made to duplicate conventional circuits which used discrete components. Examples of this type included RTL (see section 8.13) and DTL (see section 8.15). The RTL family was rapidly superseded not only by improved DTL versions, but also by other logic families. In this section of the book, a number of the more important logic families are described including

DTL	diode-transistor logic
TTL	transistor-transistor logic
ECL	emitter-coupled logic
p-MOS and CMOS	MOS logic families

A summary of the principle characteristics of the major logic families is given in section 9.9.

9.4   Diode-transistor logic (DTL)

A popular monolithic DTL NAND gate is shown in Fig. 9.7. When comparing this circuit with the discrete component version in Fig. 8.34, readers will note that diode D1 in the latter circuit is replaced by transistor TR1 in Fig. 9.7. The advantage of this arrangement is that the current flowing through the input diodes need only be equal to the base current of TR1 in Fig. 9.7; these diodes carry a much greater current in the discrete component version. This results in the monolithic version having a greater fan-out and a more rapid switching speed than is the case in the discrete component version in Fig. 8.34.

The noise immunity of Fig. 9.7 is improved if diode D2 is replaced by the Zener diode shown in inset (i) in the figure. Versions using the Zener diode have a noise immunity of about $(V_Z + 0.7)$ V, where V_Z is the breakdown voltage of the Zener diode. A disadvantage of this arrangement is that Zener diodes have a relatively large value of self-capacitance, and this reduces the switching speed of

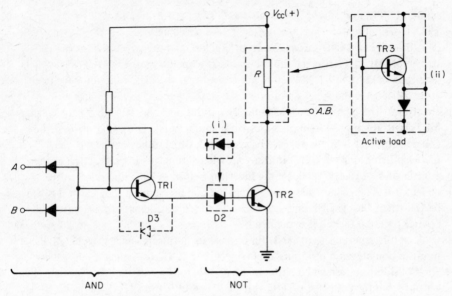

Fig. 9.7   A monolithic DTL NAND gate

the circuit. Diode D3 is sometimes used as a means of quickly discharging the charge stored in the parasitic capacitance of the Zener diode.

In this circuit (and also other types of NAND gate), unused inputs should either be connected to a logic '1' signal, or should be connected to a used input line. This provides the gate with the best possible noise immunity.

Since the output resistance of Fig. 9.7 is approximately equal to R when the output voltage is 'high', and has a small value when the output voltage is 'low', it is possible to use these gates in distributed logic or WIRED-OR networks.

The switching performance of the circuit is improved if the collector load resistance R is replaced by an active load, such as that shown in inset (ii) in Fig. 9.7. The principal reason for the improvement in switching speed is that the current gain of the transistor used in the active load reduces the effective time constant of the collector load. DTL gates with an active load of this kind cannot be used in WIRED-OR networks for the following reason. If two NAND gates with active loads have their output terminals connected together, then when TR2 is turned ON in one gate and TR3 is ON in the second gate, a short-circuit is applied to the power supply through the WIRED-OR link. This type of active load increases the 'low' output voltage by about 0·6 V.

9.5   Transistor-transistor logic (TTL)

When TTL was introduced it represented a revolution in logic circuit design, and was only made possible by monolithic IC technology. TTL has become the 'work horse' of the logic world, and is manufactured in five main forms, namely

standard TTL, low power TTL, high speed TTL, Schottky TTL and tri-state
TTL. Essential features of these types are described below.

A basic TTL NAND gate is shown in Fig. 9.8, and uses a multi-emitter
transistor (MET) TR1 in the input circuit. The operation of the MET was
described in section 9.2.2, when it was shown that the logic function developed
at its collector is the AND function of the signals applied to the emitter
electrodes. The circuit shown has only two input lines; other NAND gates having
METs with three or more inputs are also manufactured. The signal $A \cdot B$ at the
collector of TR1 is applied to the base of TR2, which functions as a
phase-splitting amplifier. By emitter following action, the logic function at the
emitter of TR2 is $A \cdot B$ and, by the inverting action of TR2, the signal at its
collector is $\overline{A \cdot B}$. Since these two signals are complementary they cause TR3 to
be ON when TR4 is OFF and vice versa. The fact that the circuit generates the
NAND function overall is explained below.

A circuit generates the NAND function if its output is logic '1' when any
input or combination of inputs are logic '0', and is '0' only when all inputs are
1's. Consider the case when either of the inputs in Fig. 9.8 are logic '0'; the
output from the collector of TR1 is logic '0', resulting in a '1' signal being
applied to the base of TR3 and in a '0' signal being applied to the base of TR4.
These respectively turn TR3 ON and TR4 OFF, so that the output line is
connected to V_{CC}, i.e., the output is logic '1'.

When $A = B = 1$, a logic '1' is applied to TR4 and a '0' to TR3. This action
causes TR4 to be in the ON state, and the output line is now connected to the

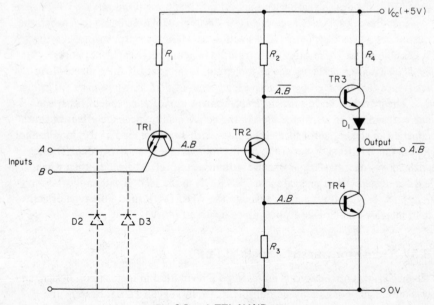

Fig. 9.8   A TTL NAND gate

zero potential line, i.e., the output is logic '0'. From the above, readers will note that the output line is coupled either to V_{CC} or to the zero voltage line via a saturated transistor and R_4, i.e., the output resistance of the circuit is low for either logic state.

Due to its appearance, the rather unusual output circuit in Fig. 9.8 is known as a *totem-pole* output circuit. The diode D1 is a voltage level shifting diode, and is included to ensure that TR3 can be turned off under all operating conditions of the circuit. The reason for resistor R_4 is explained in the following. During the turn-on and turn-off periods of operation, transistors TR3 and TR4 do not switch at the same speed. In the absence of resistor R_4, and with TR3 and TR4 conducting simultaneously, a transient short-circuit would be applied to the power supply. The function of R_4 is to restrict the peak transient current drawn from the supply at this time; in standard TTL the peak value of transient current is limited to about 40 mA per gate.

Because the output resistance of the circuit in Fig. 9.8 is low in either logic state, it cannot be used in WIRED-OR networks. This problem is overcome in one of two ways. One method is by the use of *open-collector* TTL gates, in which the components R_4, TR3 and D1 (see Fig. 9.8) are omitted. By using a single external pull-up resistor, it is possible to use a number of open-collector TTL gates in WIRED-OR networks in the same way as other logic families may be used. An alternative solution is offered by *tri-state* TTL. The latter family has a control line in addition to the normal output and input lines. The control line allows the output resistance of individual gates to be switched to a high impedance state, i.e., an open-circuited state, in which the gate will not sink more than about 40 μA. This feature allows the family to be used in WIRED-OR configurations without resorting to open-collector TTL elements.

There are three versions of the circuit shown in Fig. 9.8, namely low power TTL, standard TTL and high speed TTL, the essential difference between them being in the values of the components used in the circuits. These values are listed in Table 9.1.

The propagation time of the signal through the gate is primarily determined by the time taken for the inherent capacitances of the circuit components to be charged or discharged. To speed up this process it is necessary to reduce the circuit time constants, i.e., to reduce the values of the resistors. Hence the high speed branch of the family has the lowest values of resistance, and the largest power consumption per gate.

Table 9.1

Type	$R_1(k\Omega)$	$R_2(k\Omega)$	$R_3(k\Omega)$	$R_4(\Omega)$	Power per gate (mW)	Propagation time (ns)
Low power	40	20	12	500	1	30
Standard	4	1.6	1	130	10	12
High speed	2·8	0.75	0·5	58	20	6

Sub-units used in data processing equipment have various speed requirements, and in arithmetic sections it may be desirable to use high speed TTL. In other sections with less demanding requirements, it is possible to use standard TTL. Since the fan-out capability of a TTL gate depends on the current it has to 'sink', its fan-out varies with the type of driven gate. A list of typical fan-out values is given in Table 9.2.

Table 9.2

Driving gate	Fan-out into the following types of TTL		
	Low power	Standard	High speed
Low power	10	1	1
Standard	40	10	8
High speed	50	12	10

A feature worthy of attention in connection with TTL is the very fast rise and fall times associated with the change in output signal. In the wake of these rapid changes in signal level come problems of signal reflections along the transmitting lines. A transmission line is described as being electrically 'long' if the time taken for the signal to propagate down the line is of the same order of magnitude as the rise time of the signal being propagated. In the case of TTL, an electrically long line is one of length about 0·75–1·0 m. Under certain circumstances, the voltage reflections generated by this means may cause a transient voltage undershoot of −2 V or greater to appear on the line; this signal, when applied to the input of a driven gate, may cause damage to the gate. One solution used to overcome the worst effects of these transients is to connect diodes D2 and D3 (see Fig. 9.8) between the input lines and the common line. In normal operation each of the diodes is reverse biased (or, alternatively, has no bias applied to it), but when the input polarity is negative, the appropriate diode conducts heavily and damps out the voltage undershoot.

Small values of propagation time (≈ 3 ms) are obtained using *Schottky diode clamped TTL* gates. The general principle of reducing the storage time of logic gates by means of a clamping diode was outlined in section 8.9. In this branch of the TTL family, a Schottky diode is used as the clamping device. The Schottky diode is a rectifying metal-to-semiconductor junction device,[1] in which (in TTL) an aluminium conductor is the anode. The n-type collector region acts as the cathode. The geometry of the structure is shown in Fig. 9.9(a). The equivalent electrical circuit of the Schottky diode clamped transistor is shown in Fig. 9.9(b), and diagram (c) gives a symbol used for this structure. A complete Schottky-TTL NAND gate is illustrated in Fig. 9.9(d), the propagation time of this type of gate being about 3 ns, and the dissipation per gate is about 20 mW.

As the ambient temperature rises, so the Schottky clamp becomes less effective and the transistors are driven further into saturation. The noise

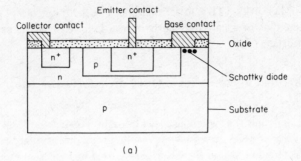

(a)

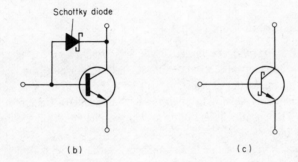

(b)   (c)

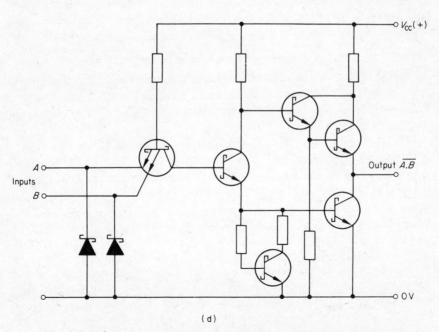

(d)

Fig. 9.9 (a), (b), (c), and (d)   Schottky diode clamped TTL

immunity of Schottky-TTL is lower than that of other branches of the TTL family. These drawbacks are frequently outweighed by the lower propagation time of Schottky-TTL.

9.6   Integrated injection logic (I^2L) or merged-transistor logic (MTL)

This logic family has the switching speed of bipolar logic together with the packing density of MOS devices. The manufacturing processes are compatible with bipolar technology and these devices are used in an increasing range of applications.

9.7   Emitter-coupled logic (ECL)

The basis of ECL is the *emitter coupled amplifier* or *long-tailed pair* circuit, which is shown in the central section of Fig. 9.10. In this circuit, the value of the current flowing in resistor R_E is constant, so that a change in the value of current (say an increase) flowing through the left-hand side of the circuit, i.e., through TR1 or TR2, results in an opposite change (a decrease) in the current flowing through TR3.

The principal advantage of this logic family over other types is that all the transistors in the circuit work in a linear or non-saturating mode. Consequently, the time delay associated with the charge storage phenomenon in saturated logic gates is eliminated. ECL is nearly always with matched line interconnections to

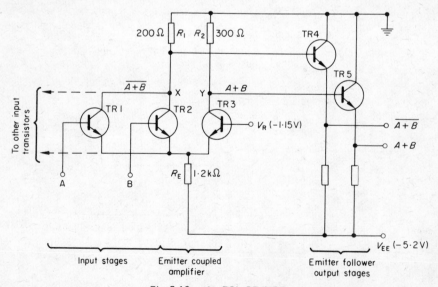

Fig. 9.10   An ECL OR/NOR gate

make full use of its high speed capability. The circuit shown simultaneously generates both the OR and the NOR functions of the input signals. Drawbacks associated with this logic family include a high power dissipation (about 60 mW per gate), and a high sensitivity to temperature change.

The operation of the circuit is now described. The base of TR3 is connected to a voltage reference source, V_R, of value about $-1 \cdot 15$ V, which lies about midway between the two logic levels. Typically, logic '0' corresponds to a voltage of about $-1 \cdot 55$ V, and logic '1' to about $-0 \cdot 75$ V, i.e., positive logic notation is used. When either input A or input B is at the logic '1' level, the current through the appropriate transistor increases, and that in TR3 decreases. As a result, the potential at point X decreases and that at Y increases. The output lines are driven by emitter follower amplifiers which fulfill two functions, namely to provide the circuit with a low output impedance and high fan-out capability, and to restore the output voltages to the correct levels for the driven stages.

Since TR1 and TR2 have their collectors connected together (the WIRED-OR connection), the logic function generated at point X is $\bar{A} \cdot \bar{B} = \overline{A + B}$. Since an increase in the voltage at X causes a corresponding reduction in the voltage at Y, then the logic function generated at point Y is $(A + B)$. By emitter follower action the logic signals at the emitters of TR4 and TR5 are the NOR and the OR functions, respectively, of the input signals.

Other names used to describe this type of circuit are *emitter-emitter coupled logic* (E^2CL), *emitter-coupled transistor logic* (ECTL) and *current-mode logic* (CML).

9.8   MOS logic gates

Basic versions of p-MOS and CMOS NOT gates were described in section 8.20. Typical p-MOS NOR and NAND gates are illustrated in diagrams (a) and (b), respectively, of Fig. 9.11. The circuits operate with a negative supply potential and use the negative logic notation. Their propagation time of about 100 ns is modest when compared with both TTL and ECL, and is limited by the input capacitances of the circuits.

The circuit in Fig. 9.12(a) is that of a 2-input CMOS NOR gate, and that in Fig. 9.12(b) is of a 2-input CMOS NAND gate. These devices use a positive supply voltage and operate in the positive logic notation. Due to the use of n-channel elements in the structure, the propagation time is reduced by a factor of about one-half[1] when compared with p-MOS logic. Moreover when the n-MOS devices conduct, the p-MOS devices in the circuit are cut off, and vice versa; since the resistance of a non-conducting MOS transistor is of the order of 5000 MΩ, the current drain per gate in either logic state is very small. The resistance of the conducting transistor is about 750 Ω, hence the output resistance for either output logic state is relatively low.

In operation, a 'high' value of input voltage causes the p-MOS device to be cut

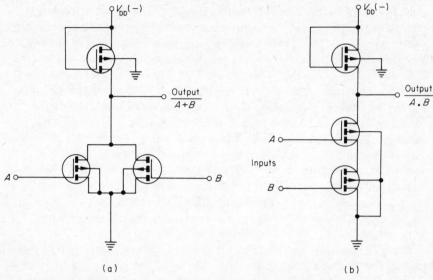

Fig. 9.11   (a) A p-MOS NOR gate, and (b) a p-MOS NAND gate

off, and the n-MOS device conducts. Only when the output logic level is changing state do both the p- and n-MOS devices conduct simultaneously for a very short period of time; however, the peak magnitude of the current pulse drawn from the supply does not usually exceed about a few hundred microamperes per gate.

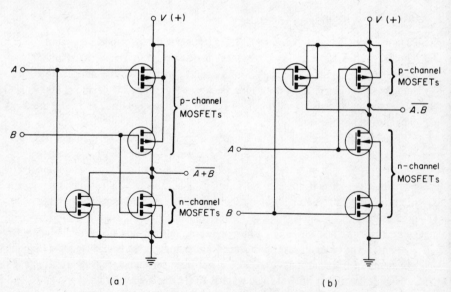

Fig. 9.12   (a) CMOS NOR gate, and (b) CMOS NAND gate

A section through a CMOS structure is illustrated in Fig. 9.13. The p-MOS devices are diffused into high-resistivity n-type material, and the n-MOS devices are formed in a 'tub' of p-type semiconductor. Additionally, heavily doped p^+ and n^+ diffusions are introduced between the FETs; these act to prevent undesirable MOSFET action in the regions between the two n-MOS devices and the two p-MOS elements, and are known as *channel stops*. If the channel stops were not introduced, unwanted inversion channels would form between the devices.

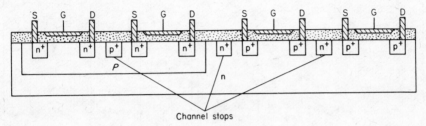

Fig. 9.13   Basic CMOS geometry

Because the gate insulation is very thin, it may easily be ruptured by the application of a comparatively low voltage such as those it may experience when being handled during manufacture and installation. To prevent damage from this cause, each input line of MOS logic families incorporates a *gate-oxide protection circuit*, a typical circuit being shown in Fig. 9.14. In the event of a transient voltage being applied to the input terminal, one or more of the diodes conduct and dissipate much of the energy in the transient pulse.

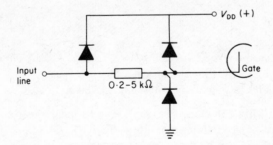

Fig. 9.14   One form of CMOS gate-oxide protection circuit

9.9   Summary of the features of the principal logic families

Table 9.3 lists some of the more important parameters of popular logic families, the figures quoted being typical of the available ranges.

The following comments also apply. TTL gates are available at low cost, and a wide variety of logic functions are manufactured in this range. They have a high

Table 9.3

Features of logic families

	Logic family				
Feature	DTL	TTL	ECL	p-MOS	CMOS
Basic function	NAND	NAND	OR/NOR	NOR	NOR or NAND
Logic notation	+	+	+	−	+
Supply voltage, V_S, volts	+5	+5	−5·2	−20	3 to 15
Logic levels '1'	3·0	3·3	−0·74	−10	$\simeq V_S$
'0'	0·2	0·2	−1·6	−2	$\simeq 0$
Noise immunity, mV	750	1000	200	1000	0·45 V_S
Fan-out	8	10	25	5	large
Propagation delay, ns	25	13	<3	100	40
Interconnection type	Current sinking	Current sinking	Current mode	Voltage	Voltage

noise immunity, together with a usefully large fan-out. The basic form of TTL gate is unable to drive 'long' lines due to problems associated with signal reflection phenomenon, and 'line driver' stages must be used in these applications.

The highest switching speeds are provided by the ECL family, whilst the lowest power consumption and most compact circuits are obtained using CMOS logic. The latter family can also work with a wide range of supply voltages, and has a large value of noise immunity.

9.10 MSI and LSI

The expressions *medium scale integration* (MSI) and *large scale integration* (LSI) are widely used to describe the complexity of certain types of logic network contained in one IC. In general their meanings are taken to be

    MSI — networks containing between about 10 and 100 gates
    LSI — networks containing more than about 100 gates.

Examples of MSI networks include digital multiplexers, shift registers and counters, whilst LSI networks include calculator chips and microprocessor chips.

9.11 MOS dynamic memories

In this family of memories, data is stored in the form of an electrical charge on the gate insulation of a MOSFET. A popular form of three-transistor circuit is shown in Fig. 9.15, in which the charge is stored in the gate-to-source

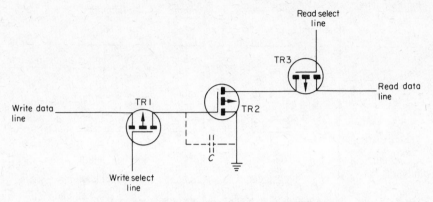

Fig. 9.15 One form of MOS dynamic memory

capacitance, C, of TR2. When a signal is applied to the 'write select line', TR1 is turned ON, and the 'write data line' is connected to the gate of TR2. Applying a logic '1' to the latter line causes capacitor C to become charged, and applying a logic '0' to the line causes C to be discharged. In this way data is 'written' into the storage location.

The data stored is sampled or 'read' by energizing the 'read select line', which turns TR3 on. The data is read by monitoring the current flowing in the 'read data line'.

The charge stored by capacitor C will ultimately decay, and it is usually necessary to *refresh* the stored charge every few milliseconds. This process is carried out automatically by means of additional logic circuitry.

9.12 Integrated circuit memory arrays

A wide variety of IC memory arrays exist, and are broadly divided into the two categories outlined in chapter 5, namely volatile stores and non-volatile stores. Included in the former are *S-R*, *J-K*, *T* and *D* flip-flops, together with dynamic memory elements of the type in section 9.11. The non-volatile category includes magnetic tape, disc and drum stores.[2] Non-volatile semiconductor memories include *read-only memories* (ROM), *programmable read-only memories* (PROM) and *reprogrammable read-only memories* (RPROM). Non-volatile semiconductor memories are used in a wide variety of applications, including storing microprograms for computers, and storing character patterns for optoelectronic display devices.

Semiconductor memory arrays are most frequently used in the form of *random access memories* (RAM), in which an individual bit or stored word can be obtained at random by means of *addressing* a location or group of locations in the store. The memory cells in a RAM are generally arranged in a matrix pattern in the form shown in Fig. 9.16. A single cell such as M3,2 may be

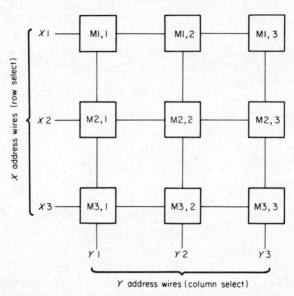

Fig. 9.16   Matrix arrangement of cells in a semiconductor RAM

addressed by energizing row wire $X3$ and column wire $Y2$. This is known as X-Y *selection* or *coincident selection*. Having selected one cell in this way, data is either written into it or is read from it by means of an additional line (or pair of lines) which run through every cell in the matrix. Non-selected cells are inhibited, and are unaffected by the signals on the data line. Alternatively, if all the 'row select' wires and the $Y2$ wire are addressed simultaneously, it causes memories M1,2, M2,2 and M3,2 to be addressed at the same time. This is known as *word selection* or *linear selection*. Data can simultaneously either be read from or written into the selected group of cells.

In some cases, it is preferable to address a memory by means of its stored contents, rather than by its address within a storage bank. For example, it is sometimes convenient in an educational establishment to store data about a student in the form of the course he attends, the day or days he attends, his age and the name of his employer. This information can be stored in a *content addressable memory* (CAM) in which the stored data is addressed simply by inspecting part of the contents of the memory. Thus, it would quickly be possible to locate, for example, students who are employed by employer X, and who are attending course Y on day Z.

9.13   ROMs, PROMs, and RPROMs

The name read-only memory (ROM) is applied to memories whose stored data cannot be altered. Many of these fall into the category of *mask programmed ROMs*, which contain data in the form of 1's and 0's at addresses specified by the user; the data is written into the memory during the IC manufacturing process by means of photographic masks, and these specify which diodes or transistors in the array are in the ON state and those which are in the OFF state.

A small part of one form of *electrically programmable ROM* (PROM) is shown in Fig. 9.17; in this case each memory element consists of a diode in series with a fusible link. The fuses can be 'blown' individually by addressing a location and applying a current impulse to it. When the memory is to be left ON, the fusible link is left intact, and where it is to be OFF the fuse is blown. The programming is carried out by means of electronic apparatus, and can be completed either by the manufacturer or by the user.

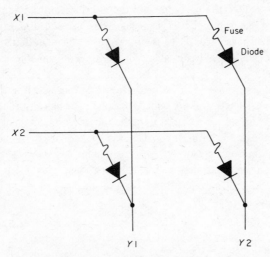

Fig. 9.17   One form of PROM

Reprogrammable read only memories (RPROM) contain MOS devices into which it is possible to introduce a 'permanent' charge package in the gate oxide regions. In this type of device the stored charge does, in fact, slowly leak away, but it may take more than about 10^{11} read accesses for the charge to have reduced to an unusable value. The data stored is usually erased by exposing the chip to ultraviolet radiation. After erasure, new data can be electrically written into the memory, the 'writing' period usually taking a few milliseconds. Since RPROMs are more frequently used as ROMs, they are also known as *read-mostly memories* (RMM).

When a ROM has been programmed, it can be regarded as a logic array which contains a number of combinational sequences, different combinations being obtained by addressing combinations of input (address) lines. In this way the ROM can be used to replace a number of combinational logic arrays. When used in this mode, it is described as a *programmable logic array* (PLA).

9.14   Microprocessors

A microprocessor can simply be thought of as the basic elements of a computer contained either on a single IC chip or on only two or three chips. The microprocessor contains three basic sections, namely:

1. The arithmetic and logic unit (ALU) or central processor
2. A memory bank
3. An interface unit

The central processor carries out the mathematical operations such as addition, subtraction, multiplication and division, as well as shifting data either 'right' or 'left', incrementing, decrementing, etc. The memory is used to store data and instructions relevant to the problem, and the interface unit handles incoming and outgoing signals. In practice, it is necessary to use additional ICs to supplement the storage capacity of the microprocessor, together with other elements such as analogue-to-digital and digital-to-analogue convertors which may be required.

Microprocessors promise to revolutionize the way in which both systems and instruments are designed, and are widely used in information processing systems. Since an instruction set can be used to make the microprocessor take on the appearance of a logic function, it can also be used as a programmable logic array. In the design of microprocessor systems, experience has shown that about 8—16 memory locations are equivalent to one gate in a logic array. With an 'average' complexity in a logic IC, about 80—160 memory cells are needed to replace the equivalent of one IC package.

9.15   Charge-coupled devices

A charge-coupled device (CCD) is a multi-gate MOS element, which can be caused to transfer charge 'packages' between its source and drain under the control of multi-phase gate pulses.[1,3,4]

The operation of the charge transfer mechanism can be understood by reference to Fig. 9.18. In the figure, a p-MOS structure is used, and the application of a negative potential to a CCD electrode produces a depletion layer in the substrate beneath it, the greater the electrode potential the greater the depth of the depletion layer. When the gate voltage exceeds the threshold voltage of the device, a local inversion channel is developed beneath the gate electrode. The basic CCD element in Fig. 9.18 has three electrodes, each being energized by one phase of a three phase supply. The three phase supply voltages are designated the symbols ϕ_1, ϕ_2 and ϕ_3. Initially, potential ϕ_1 has the greatest value (see diagram (a)), so that a potential 'well' is formed below this electrode. Also, the holes in the local inversion channel concentrate below electrode ϕ_1. The operation can be understood more easily if the holes in the inversion channel are thought of as 'filling' the lower part of the potential well. This

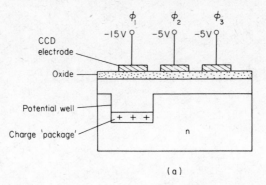

(a)

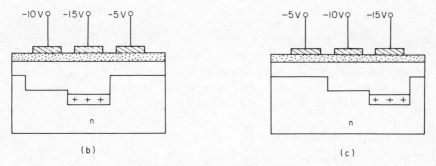

(b)   (c)

Fig. 9.18 (a), (b), and (c)   A CCD element

charge packet is transferred from electrode ϕ_1 to ϕ_2, and thence to ϕ_3 by applying the sequence of voltages shown in diagrams (a) to (c) of Fig. 9.18.

A shift register[4] is formed by cascading a number of CCD elements, the data either being injected in a serial mode or in a parallel mode. The CCD shifting principle finds a range of applications including shift registers, analogue delay lines and optical imaging.

References

1. MORRIS, N. M., *Electronic Devices*, Macmillan
2. WOOLONS, D. J., *Introduction to Digital Computer Design*, McGraw-Hill.
3. BOYLE, W. S., and G. E. SMITH, 'Charge-coupled semiconductor devices', *Bell Syst. Tech. J.,* **49**, 587–93, 1970
4. TOMPSETT, M. F., G. F. AMELIO, and G. E. SMITH, 'Charge-coupled 8-bit shift register', *Appl. Phys. Letters,* **17**, 111–115, 1970

10. Alphanumeric displays

An expanding digital instrument and digital calculator market has lead to the manufacture of a wide range of alphanumeric display devices. These include light emitting diodes (LED), liquid crystal displays (LCD), planar gas-discharge displays, planar filament displays and phosphor diodes.

10.1 Light emitting diodes

LEDs are semiconductor p-n junction devices which emit optical radiation when forward biased. The wavelength of the radiated energy depends on the forbidden band-gap of the semiconductor; the most useful range of semiconductor materials include those based on gallium arsenide and gallium arsenide phosphide. Using these materials, a range of colours can be generated, the most popular being red, green, yellow and amber. The life expectancy of these devices is typically greater than 10^5 hours and, in general, they do not fail unpredictably but degrade slowly. The power supply requirements depend on the material used in the construction of the diode (and therefore on the colour radiated), and are typically 2–2·5 V at 5–40 mA. These values of p.d. and current are known as the *forward voltage*, V_F, and the *forward current*, I_F, respectively. The reverse breakdown voltage of LEDs is typically 3–10 V. When used with a.c. supplies it is usual to protect the diode from reverse breakdown by connecting a conventional p-n junction diode either in series with or in inverse-parallel with the LED; these modifications are illustrated in insets (i) and (ii), respectively, in Fig. 10.1. When the LED is conducting, the p.d. across it falls to V_F; the current through the diode is limited in value to a safe value, I_F, by resistor R, whose value is estimated from the equation

$$R = (V_S - V_F)/I_F$$

where V_S is the value of the supply voltage. This resistor, in the case of a conventional LED, is a discrete component which is connected externally to the

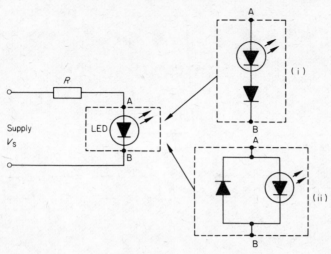

Fig. 10.1   Basic LED circuit

LED. Certain types of LEDs, known as *resistor LEDs*, contain an integral current limiting resistor, and can be connected directly to the voltage specified for the diode; this voltage may, for example, be 5 V for use with TTL networks.

LEDs may either be used individually in the form of indicator lamps, or may be used in groups to provide either numerical or alphabetical character displays, or a combination of the two types of display (*alphanumeric displays*).

SEGMENTED DISPLAYS

The *seven-segment display* in Fig. 10.2(a) is widely used in numerical display systems, such as those found in instruments and in calculators. By illuminating combinations of segments, it is possible either to generate any one of the ten decimal digits (see Fig. 10.2(b)), or to generate one of a limited range of alphabetical characters (see Fig. 10.2(c)).

A block diagram of a 7-segment display system is illustrated in Fig. 10.3. The input pulses to be counted and displayed are applied serially to the system. These pulses are counted in a decade counter, the final value of the count being transferred to a 'data latch' unit containing D flip-flops. The function of the data latch is to provide a 'non-blinking' display in the time interval during which the incoming pulses are being counted by the counter module. When the counting sequence has been completed a 'latch strobe' pulse is generated by the logic circuitry, which causes the logic states in the counter to be transferred to the data latch element. If the incoming data is presented in a parallel form, then the decade counter module is redundant; if a 'non-blinking' display is not required, then the data latch is also surplus to requirements.

The signals on the output lines of the data latch are applied to a decoder, which provides signals on seven output lines which activate the segments of the

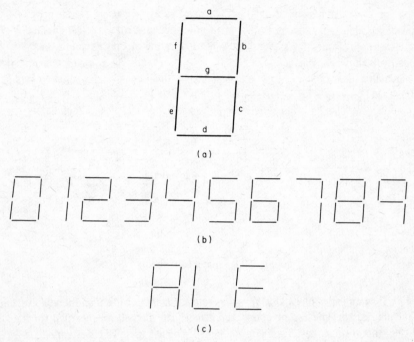

(a)

(b)

(c)

Fig. 10.2 (a), (b), and (c)   The seven-segment display

7-segment display. The design of a decoder for this purpose can be carried out using the principles outlined in chapter seven. To provide the LEDs in the display with the correct value of current, driver circuits are interposed between the decoder and the display elements. The LED driver block in Fig. 10.3 has an additional input marked 'blanking input'; when a signal is applied to this line, it causes the display to be completely blanked out, i.e., it is not illuminated. This

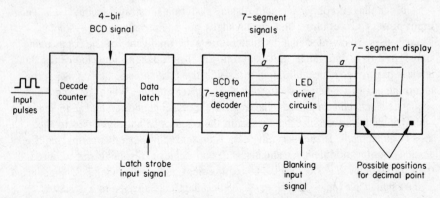

Fig. 10.3   A block diagram of a 7-segment display system

facility is used to blank out non-significant zeros in the display. In addition to the seven segments used to display both numerical and alphabetical characters, decimal point indication can also be provided in one of the two alternative positions shown in Fig. 10.3.

An alternative form of segmented display is the 16-segment pattern in Fig. 10.4(a). This type of display can generate all alphabetic and decimal characters, in addition to which other patterns can be generated. Several of the characters which can be displayed are illustrated in Fig. 10.4(b). Displays of this kind are frequently generated via a ROM character generator.

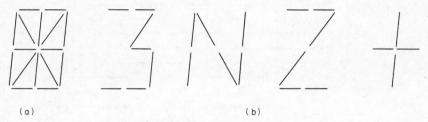

(a)   (b)

Fig. 10.4 (a) and (b)   A 16-segment display

MATRIX DISPLAYS

Multi-dot matrix displays are widely used, the 5 x 7 dot matrix display (Fig. 10.5(a)) being very popular; the 4 x 7 dot matrix display in Fig. 10.5(b) being an alternative type. Characters specified by the ASCII code (American Standard Code for Information Interface) can be displayed on a 5 x 7 alphanumeric matrix display. The 5 x 7 display in Fig. 10.5(c) is a simplified version of that in diagram (a), and is capable of generating the ten decimal digits together with a limited range of alphabetical characters. Physically large displays can be produced using a matrix of LED lamps.

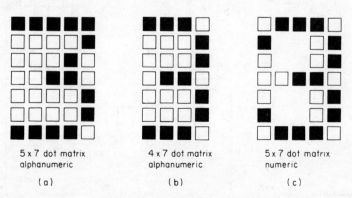

5 x 7 dot matrix   4 x 7 dot matrix   5 x 7 dot matrix
alphanumeric   alphanumeric   numeric

(a)   (b)   (c)

Fig. 10.5   Dot matrix displays: (a) 5 x 7 alphanumeric, (b) 4 x 7 alphanumeric,
(c) 5 x 7 numeric

MULTIPLE DISPLAY DEVICES

To economize on the number of drive circuits required to display a large number of display devices within one instrument, a method known as *scanning* or *strobing* or *multiplexing* is used.

The basis of one form of scanning system is shown in Fig. 10.6. The character to be displayed and the LED array on which it is to be displayed are specified by signals on the incoming data lines. This information is stored in buffer memories which are addressed sequentially by the logic system. To form a character, one

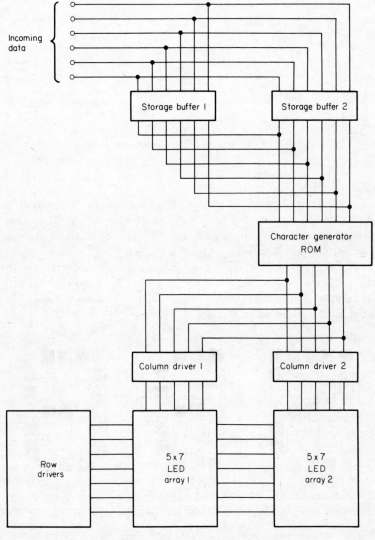

Fig. 10.6   The basis of a multi-LED scanning circuit

row of diodes at a time is selected or is addressed by the row driver circuit, commencing with the upper row. When the upper row is selected, the data in storage buffer 1 is applied to the character generator ROM, which causes the correct signals to be fed to the LEDs in the upper row of array 1 via the column drivers. After scanning the first row, the second row of diodes are selected by the row drivers, and the appropriate LEDs are illuminated. After all the rows have been addressed one at a time, the cycle is repeated. Individual LED arrays are addressed sequentially, and in this way one ROM is shared between all the displays. The basic system shown can be extended to deal with a large number of LED arrays; a flicker-free display is obtained by scanning each complete character at least 100 times per second, so that the scanning frequency for a 10 character display should be at least 1 kHz. However, in order to produce the same level of illumination as LEDs energized by a constant voltage, LEDs in scanning systems must be energized by a much higher value of current. Tests on LEDs operating under pulsed conditions show that they have a lifetime expectancy exceeding 10^6 mAh.

10.2   Liquid crystal displays (LCD)

Liquid crystals are organic fluids existing in a mesophase between their solid and liquid states. Unlike LEDs, LCDs do not radiate illumination, but either reflect or transmit incident illumination. In operation, the power requirement of a LCD is minute, being typically 15 μW or less for a 5 mm (0·2 in) high seven-segment display. If the source of illumination is ambient lighting, then no additional power supply is required, but if the ambient illumination is zero, then a light source is required. Since the display brilliance of LCDs is a function of the incident illumination, direct lighting improves the display. In general, where a principal requirement is either a physically large display together with a low power consumption, then LCDs are superior to other types.

There are many hundreds of liquid crystal compounds, which are divided into three groups, namely *nematic, smetic* and *cholestric*, the former being of most interest here.

There are two types of nematic liquid crystal displays: one is the *dynamic scattering* type which causes light to be reflected, the other being known as *field effect* or *twisted nematic* type which allows light to be transmitted. Dynamic scattering displays were introduced in 1967, whilst field effect displays were introduced in 1970.

The general form of construction of a 7-segment display is illustrated in Fig. 10.7(a), the liquid crystal being sealed between two glass surfaces (one which may have a mirror finish), the surfaces having transparent conductive coatings on them. A typical spacing between the front and back plates is 10 μm. Any shape of display can be obtained using this method, and diagrams (b) and (c) are for a 7-segment numerical indicator. The lower electrode (b) is common to all

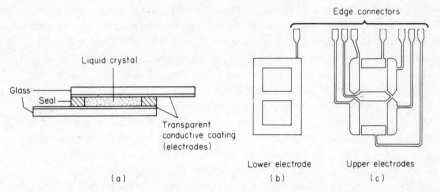

Fig. 10.7 Liquid crystal display: (a) section through the display, (b) the lower electrode for a 7-segment display, (c) one form of arrangement of the upper electrodes

segments, whilst the upper segments (c) are energized independently; the two sets of electrodes are aligned above one another in the display unit.

In order to produce the desired display, an electric field of about 0·5—1 MV/m (say 5—10 V across a 10 μm film) is applied to the liquid crystal. This causes the crystal molecules to rearrange their orientation (see below). There is very little spreading of the re-orientation beyond the edges of the electrodes.

Electrolyte ions are an essential ingredient of liquid crystals and, if the display is energized by a d.c. supply, there is a tendency for electrolytic dissociation of the liquid crystal. This dissociation leads to a low deterioration in the display; LCDs do not fail catastrophically from this cause. Considerably extended lifetimes (>10 000 h) are obtained by exciting the LCD by a.c. signals in the frequency range 30—100 Hz. A typical LCD drive circuit for a single segment is shown in Fig. 10.8(a); in this circuit the common electrode (A) is connected to the clock source, whilst the 'displayed' electrode is driven by the EXCLUSIVE-OR gate G1. When the signal on the 'select' line is logic '0', the p.d. across the LCD element is zero and the display remains transparent. When a logic '1' is applied to the 'select' line, an a.c. signal is applied across the element, and it becomes visible. Typical waveforms in the circuit are illustrated in Fig. 10.8. One EXCLUSIVE-OR gate per segment is required. A list of the more important characteristics of the two types of LCD are given in Table 10.1.

DYNAMIC SCATTERING LCDs

The molecules in dynamic scattering displays are thread-like in form, and in the absence of an electric field the molecules align with one another under the influence of the molecular forces. In the unenergized state, light is transmitted through the liquid crystal and the display is transparent.

The application of a potential across the LCD causes turbulence domains to form within the film, and the molecules beneath the energized segments in these

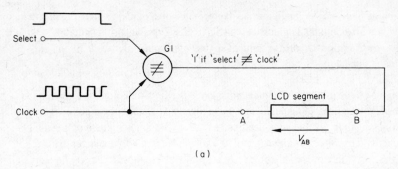

(a)

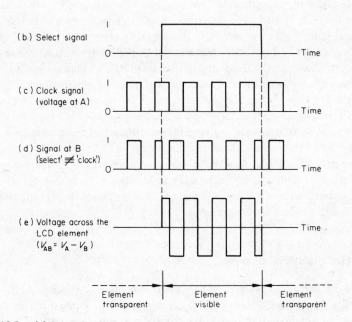

Fig. 10.8   (a) Drive circuit for an LCD element, and (b) to (e) show typical waveforms

Table 10.1

Characteristics of LCDs

	Dynamic scattering	Field effect
Voltage	15—25	1·5—10
Current/cm^2	1 μA	0·4 μA
Viewing angle	160°	90°—120°
Switching time	0·3 s	0·1—0·3 s
Life	10 000 h	10 000 h
Temperature range	0—80°C	0—70°C

domains become very efficient scatterers of white light, causing the segment to have a white colour. The contrast ratio of this type of display increases with the value of the applied voltage, and rises to a maximum of about 20 : 1.

FIELD EFFECT LCDs

In field effect or twisted nematic displays, the liquid crystal is sandwiched between two polarizing surfaces whose planes of polarization are at right angles to one another. In the unexcited state, the molecular structure of the LCD causes the plane of polarization of the incident light to be twisted through 90°, so that it is transmitted through the second polarizing surface. If the lower glass surface has a mirror finish, the light is reflected back through the liquid crystal, where the plane of polarization is given a reverse 90° phase shift, so that it arrives at the upper polarizer in the correct plane to pass through it. In the unexcited mode, the display is therefore completely transparent.

The application of a potential between the LCD electrodes causes the molecules to align, and the incident light is transmitted without its plane of polarization being twisted. In this mode of operation, the planes of polarization of the light and that of the lower polarizer are at 90° to one another, and the incident light is completely absorbed by the lower polarizer. The result is a dark character which contrasts with the brighter surroundings. Contrast ratios in the range 25 : 1 to 35 : 1 are obtained from field effect LCDs.

Transmittive field effect displays do not have an internal mirror, and transmit light rather than reflect it. Hence it is possible either to have displays which have dark characters on a light background, or light characters on a dark background.

10.3   Gas discharge displays

Gas-filled discharge displays were widely used in many early types of digital instruments. These devices contained ten cathodes and an anode in a gas-filled glass tube; each cathode is shaped in the form of a decimal number and, when excited, glows a characteristic pink colour. These tubes require a drive voltage in the range 160–200 V, which is ideal for mains equipment but is not quite so suitable for portable apparatus. These displays are often difficult to read in conditions of high ambient illumination and, when viewed at angle, the display appears to 'dance' as it changes. Moreover, due to the nature of the construction of the tube, the viewing angle is limited.

Planar 7 x 5 dot matrix gas discharge displays overcome the problem of display 'dancing' and of the small viewing angle. Moreover, matrix displays can be multiplexed, so that circuit simplification is possible. Planar 7-segment gas discharge displays are also used in measuring instruments and in calculator displays.

10.4 Filament displays

Planar 7-segment filament displays are used in instruments and in industrial applications, and can be driven from logic systems. Typically, a display of height 16—20 mm (0·4—0·6 in) can be operated from voltages in the range 4—6 V at a current of about 10 mA. Larger displays require a higher operating voltage. A typical filament life is about 10 000 h.

10.5 Phosphor diode displays

Planar 7-segment phosphor diode (fluorescent) displays are widely used in calculators and in electronic instruments. They operate on the 'magic eye' tuning indicator principle, and employ anode plates coated with a phosphor which glows a characteristic blue-green colour when bombarded with electrons. The anode plates are shaped in the form of the seven segments of the display; the anode voltage is about 20 V, and the current per segment is typically 1—3 mA. A heater supply of about 1·5—5 V is also required for each indicator tube.

11. Digital-to-analogue and analogue-to-digital convertors

Analogue control systems used in industry frequently have input signals derived from digital sources such as paper tapes, magnetic tapes, computers, etc., and a digital-to-analogue convertor (DAC) is an essential part of the interface between the digital and analogue elements of the system. Conversely, digital systems and data loggers must be capable of accepting analogue signals from transducers and instruments. In this case, there is a need for analogue-to-digital convertors (ADC) in the interface units.

11.1 Digital-to-analogue convertors

DACs usually consist of an arrangement of resistors of carefully weighted values, and are used to convert signals corresponding to weighted binary codes into an equivalent analogue voltage.

One basic network is the star-connected circuit of weighted resistors in Fig. 11.1. Switches A to D can be either mechanical contacts or transistor switches. With the switches in the positions shown, the output (V_0) from the circuit is zero. In the circuit shown, the resistors are weighted for the 8421 BCD code, the most significant bit being applied to the resistor with the lowest value. The output voltage is given by the expression

$$V_0 = V_{ref}(8A + 4B + 2C + D)/8R \left(\frac{1}{R_L} + \frac{15}{8R} \right)$$

In many cases, $1/R_L \ll 15/8R$, when

$$V_0 = V_{ref}(8A + 4B + 2C + D)/15$$

Thus, when $A = 1$ it contributes an output voltage of $8V_{ref}/15 = 0·533V_{ref}$, and when $B = 1$ it contributes an output of $0·267V_{ref}$, etc. The accuracy of this type

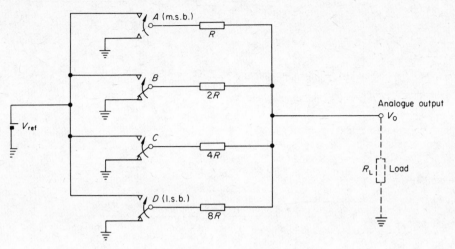

Fig. 11.1   Star-connected summation circuit

of convertor depends on the *relative* values of the resistors rather than on their *absolute* values, and resistor networks for DACs are readily available in the form of film circuits. The DAC in Fig. 11.1 also requires an accurate and stable voltage reference source. This type of DAC can be extended to deal with decades of BCD values by using higher resistance values. For example, the most significant decade would use resistors of values R, $2R$, $4R$, and $8R$, whilst the next lower significant decade would use resistor values of $10R$, $20R$, $40R$ and $80R$, etc.

The digital-to-analogue *conversion time* is limited by the time taken for the least significant bit to propagate along its input resistor to the summing junction (this bit takes the longest time of all the input signals, since it is associated with the largest value of resistance). The signal can usually be assumed to have fully propagated after a time of about $5RC$ seconds, where R is the resistance (in ohms) of the l.s.b. input resistance and C is the self-capacitance (in farads) at the summing junction. When computing the value of the settling time, readers should be aware that it may be necessary to account for the resistance loading connected between the summing junction and the common line. This effectively increases the time constant of the conversion circuit.

The values of the resistors used depend on the 'weighting' of the code being converted. A 2421 BCD code would use resistors weighted in the order $2R$, R, $2R$, and $4R$, respectively.

Another very popular type of DAC is the R-$2R$ ladder network in Fig. 11.2. This type of network uses only two values of resistance, and is readily available in film integrated circuit form. As before, the conversion time is limited by the time taken for the l.s.b. to propagate along the ladder. However, settling times of

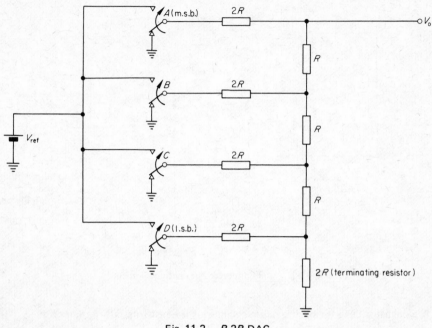

Fig. 11.2   *R*-2*R* DAC

the order of 40 ns are obtainable. The output voltage from the circuit in Fig.
11.2 is given by the expression

$$V_0 = V_{ref}(8A + 4B + 2C + D)/16$$

where A, B, C and D have either unity or zero logical value, corresponding to the
positions of the switches.

11.2   Continuous balance ADC

The basis of the continuous balance ADC is illustrated by means of the circuit in
Fig. 11.3(a). An unknown value of analogue voltage, V_X, is applied to a
comparator together with a second voltage V_Y. The latter voltage is obtained
from a counter via a DAC of the type described in section 11.1; the value of V_Y
is therefore proportional to the number stored in the counter. At the
commencement of operations, the states of the counter are reset to zero, and the
initial value of V_Y is zero. During the time interval when $V_X > V_Y$, the output
signal from the comparator is logic '1'; this signal opens the AND gate and allows
pulses to be applied to the counter. So long as the gate remains open, pulses are
counted and V_Y continues to increase in the manner illustrated in Fig. 11.3(b).
If, for example, a voltage of 1·1 V is to be measured and each clock pulse causes
the output of the DAC to increase by 0·1 V, then the value of V_Y reaches 1·1 V
when eleven clock pulses have been counted. When $V_Y \geqslant V_X$, the comparator

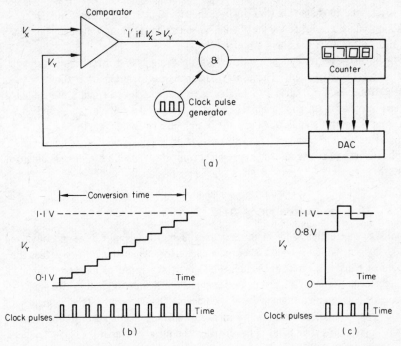

Fig. 11.3   (a) The basis of a continuous balance ADC, and (b) shows how an input of 1·1 V may be encoded. Diagram (c) illustrates how 1·1 V is encoded using a successive approximation technique

output falls to zero and no further pulses are transmitted to the counter. By means of a suitable scaling factor, the value stored in the counter is displayed in terms of the value of V_X. By using a reversible counter, the ADC can 'track' variations in the value of the applied voltage.

This type of counter is relatively simple to construct, but the conversion time may be relatively long if a large value of voltage is to be measured.

The *successive approximation ADC* is a variation of the continuous balance type, the difference between the two being in the sequence of trial voltages used to obtain the final balance. The successive approximation ADC commences by incrementing the most significant bit, and then progressively either adds or discards less significant bits until balance is reached. Assuming that the counter operates the 8421 BCD code, and that the least significant '1' causes the DAC to generate 0·1 V, then the following sequence of events occurs when balancing an input of 1·1 V. This sequence of events is illustrated in Fig. 11.3(c).

8 (0·8 V) tried and retained
4 (0·4 V) tried and rejected
2 (0·2 V) tried and retained
1 (0·1 V) tried and retained.

As illustrated in the above description, the successive approximation ADC reaches balance in far fewer steps than does the continuous balance type, but requires more complex logic circuitry.

Both of the above types of convertors are capable of converting voltages to an accuracy of 0·01 percent. They suffer from the disadvantage that their speed of conversion is so fast that they can give false readings if electrical noise is superimposed on the input signal. The effects of noise voltage can be reduced by filtering the input signal, but only at the expense of increasing the data conversion time. The successive approximation type is widely used in applications where speed of conversion is important, such as in data acquisition systems.

11.3   Voltage-to-frequency ADC

In voltage-to-frequency (V-f) convertors, the analogue input signal is converted into a current which is used to charge a capacitor. When the voltage across the capacitor reaches a threshold value, the capacitor is rapidly discharged through an electronic device. This process is repeated continuously, and in this way a sawtooth voltage waveform is developed across the capacitor. The frequency of this waveform is measured, and by means of suitable scaling factors the frequency reading is calibrated in terms of the input voltage.

The waveform generated is gated into the frequency meter by means of a clock pulse, the pulse period being selected so as to minimize the worst effects of interference at mains frequency or at some other frequency. However, the accuracy is still dependent on a number of factors including the accuracy of the voltage-to-charging current element, the stability of the capacitance of the capacitor, and on the accuracy of determining the threshold voltage at which the capacitor is discharged.

V-f ADCs are also manufactured using feedback techniques, in which a precision frequency-to-voltage convertor is used in the feedback path.

In many convertors, an input voltage of zero causes the V-f oscillator to generate a frequency which is in the centre of its range. If this is 200 kHz, then input signals of positive polarity cause the frequency to increase above 200 kHz, and negative polarity signals produce frequencies which are below the 'centre' value. In this way it is possible to detect and indicate a reversal of polarity.

11.4   Dual-slope or dual-ramp ADC

The basis of the dual-slope ADC is illustrated in Fig. 11.4(a). The circuit shown is only suitable for measuring voltages with a positive polarity, but negative potentials can be measured by means of a relatively simple modification.

The incoming voltage, V_X, is first connected to a phase inverting electronic integrator. Since V_X has a positive polarity, then V_Y has a negative polarity which increases at the rate of V_X/RC V/s (see curve A in Fig. 11.4(b)). So long

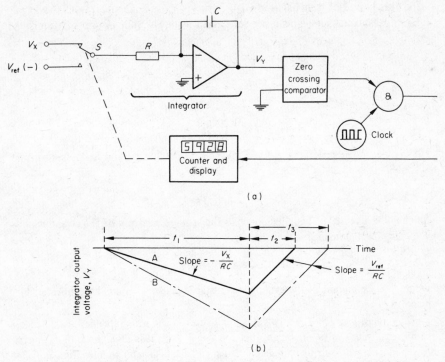

Fig. 11.4   (a) The basis of a dual-slope ADC, and (b) waveforms at the output of the integrator for a constant value of incoming voltage, V_X

as V_Y is non-zero, the zero crossing comparator element produces an output of logic '1'. This signal opens the AND gate and allows clock pulses to be applied to the counter. At the commencement of the sequence, all the stages in the counter are reset to zero; the time interval, t_1 (see Fig. 11.4(b)), during which voltage V_X charges capacitor C is determined by the time taken for the counter to reach its maximum capacity. With a four digit counter, this is the time taken for it to count 9999 pulses. The 10 000th pulse causes all the stages of the counter to be reset to zero, and the resulting 'carry' pulse causes the blade of switch S to be connected to the reference voltage, V_{ref}, which has a negative polarity. The phase inverting action of the integrator causes its output voltage to rise in a positive direction and, after time interval t_2 (see Fig. 11.4(b)), the integrator output voltage becomes zero again. At this instant of time the output from the comparator falls to logic '0', and prevents any further clock pulses being applied to the counter. As shown below, the number stored in the counter is proportional to the value of V_X.

Referring to Fig. 11.4(b), readers will note that after a length of time t_1 the integrator output voltage is

$$V_Y = -V_X t_1 / RC \text{ volts}$$

After this period of time has elapsed, V_{ref} is applied to the integrator. The integrator output voltage becomes zero again when the following condition is satisfied.

$$\frac{V_X t_1}{RC} = \frac{V_{ref} t_2}{RC}$$

or when

$$V_X = V_{ref} t_2 / t_1$$

If the clock pulse generator produces N_1 pulses in time t_1, and N_2 pulses in time t_2, then

$$V_X = V_{ref} N_2 / N_1$$

Since N_1 is a constant which is equal to the storage capacity of the counter, and V_{ref} also has a constant value, then

$$V_X = k N_2$$

where k is a constant of the system. By suitable scaling, N_2 can be calibrated directly in terms of V_X.

Curve B in Fig. 11.4(b) illustrates the effect on the integrator output voltage of increasing the value of the applied voltage, V_X. In this case, the time t_3 taken for the integrator output voltage to return to zero is greater than in the case of curve A. Consequently, the number stored in the counter is greater in the case of curve B than for curve A.

The dual-slope technique can be implemented at a lower cost than can any other method, and its use is widespread in low-cost digital instruments. It is also well suited to fabrication in IC form. Also, since the clock pulse rate is used as a means of measuring both the charge and discharge periods of the capacitor, the accuracy of measurement is immune to long-term variations in clock frequency or in changes in the time constant of the integrator. Moreover, since the circuit contains an integrator, the net effects of transient inputs are slight.

11.5   Digital frequency synthesiser

A circuit which is the frequency domain analogue of the DAC is the digital frequency synthesiser in Fig. 11.5. A basic section of the circuit is the phase comparator and averaging section in Fig. 11.5(a). In this section of the circuit, a signal of frequency f_{ref} is applied to the S-input of an S-R flip-flop, and another signal of frequency f_2 is applied to the R-input. In operation, the two signals have equal values of frequency, but differ in phase from one another; the flip-flop generates a square wave whose output pulse width is proportional to the phase difference $(\phi_{ref} - \phi_2)$, between the two signals. The low-pass filter removes the a.c. component of the waveform, and the output from the circuit in

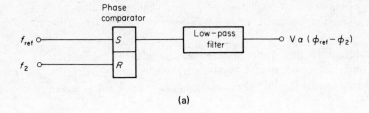

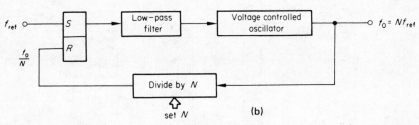

Fig. 11.5 (a) and (b)   The basis of a frequency synthesiser using the phase-lock loop technique

Fig. 11.5(a) is a d.c. potential whose magnitude is proportional to the phase difference between the two signals applied to the flip-flop.

A block diagram of the frequency synthesiser is shown in Fig. 11.5(b). The voltage appearing at the output of the low-pass filter is applied to a *voltage controlled oscillator* (or V-f convertor), the output from this being fed back to the R-input of the flip-flop via a divide-by-N frequency divider; the frequency applied to the R-input is therefore f_0/N. Under steady-state operating conditions,

$$f_{ref} = f_0/N$$

or

$$f_0 = Nf_{ref}$$

Hence, the circuit produces an output frequency which is an integral multiple of the reference frequency. The technique described above is known as the *phase-lock loop technique*. The value of the multiplication factor N can either be set by means of manual switches, or by digital signals.

This method of multiplication is widely used in signal generators and in communications systems.

Appendix

Graphic symbols used for logic functions are many and varied, and for this reason simple, although non-standard, symbols are used in this book. To enable readers to interpret logic diagrams using standard symbols, a condensed list is given in Table A.1.

Table A.1

Logic function	Symbol		
	British	American	European
Invertor			
OR			
NOR			
AND			
NAND			
EXCLUSIVE – OR			
S – R flip – flop or bistable element	S Q R $\bar{Q}$	S Q R $\bar{Q}$	S Q R $\bar{Q}$
T flip – flop or toggle flip – flop	T Q $\bar{Q}$	T Q $\bar{Q}$	T Q $\bar{Q}$
J – K flip – flop	J Clock K Q $\bar{Q}$	J Clock K Q $\bar{Q}$	J Clock K Q $\bar{Q}$
D flip – flop	D Clock Q $\bar{Q}$	D Clock Q $\bar{Q}$	D Clock Q $\bar{Q}$

214

Bibliography

BUDINSKY, J., *Techniques of transistor switching circuits*, Iliffe.

CALDWELL, S. H., *Switching circuits and logical design*, John Wiley.

CLEARY, J. F., (Ed.), *General Electric transistor manual*, International General Electric Company.

DOKTER, F. and J. STEINHALIR, *Digital Electronics*, Macmillan.

FLEGG, H. C., *Boolean algebra*, Blackie.

FOSTER, K. and G. A. PARKER, *Fluidics, Components and Circuits*, John Wiley.

HOFFMANN de VISME, G. F. A., *Binary Sequences*, E.U.P.

LEWIN, D., *Logical Design of Switching Circuits*, Nelson.

MALEY, G. A. and J. EARLE, *The logical design of transistor digital computers*, Prentice-Hall.

MILLMAN, J. and C. C. HALKIAS, *Integrated Electronics*, McGraw-Hill.

MORRIS, N. M., *Advanced Industrial Electronics*, McGraw-Hill.

MORRIS, N. M., *Semiconductor Devices*, Macmillan.

WOOLONS, D. J., *Introduction to Digital Computer Design*, McGraw-Hill.

Solutions to numerical problems

Digits marked thus ⌐‾‾‾¬ are repeated terms.

Chapter 1
1.2. (a) 1010010010100,   (b) 10111·11,   (c) 0·00000̅0̅1̅1̅
1.3. (a) 5798,   (b) 1500,   (c) 105,   (d) 111
1.4. (a) 0·0001,   (b) 0·111,   (c) 0·10̅1̅0̅,   (d) 0·10001
1.5. (a) 15035,   (b) 0·4375,   (c) 23·25
1.6. (a) 1735,   (b) 3220,   (c) 12001,   (d) 1021112
1.7. (a) 3124,   (b) 111001,   (c) 464,   (d) 2022
1.11. (a) 0101,0010,0100,1001,   (b) 1000,0011,0111,1111
1.12. 25
1.13. Odd parity. Error in second bit, second word. 9257.

Chapter 2
2.1. (a) 1011,   (b) 100100,   (c) 11110·01,   (d) 111·1011,
    (e) 10010·101
2.2. (a) 11,   (b) 1010,   (c) −1,   (d) 10,   (e) 1·11
2.3. (a) 10010,   (b) 100011,   (c) 11100·001,   (d) 1·0001,
    (e) 0·1̅0̅0̅1̅,   (f) −1100
2.4. (a) 10·0,   (b) 11·0,   (c) 0·0011,   (d) 1010,
    (e) 110·1

Index

Printed in Great Britain by J. W. Arrowsmith Ltd., Bristol